Advances in Textile Structural Composites II

Advances in Textile Structural Composites II

Guest Editor

Rajesh Kumar Mishra

MDPI

Basel • Beijing • Wuhan • Barcelona • Belgrade • Novi Sad • Cluj • Manchester

Guest Editor
Rajesh Kumar Mishra
Department of Material
Science and Manufacturing
Technology
Czech University of Life
Sciences Prague
Prague
Czech Republic

Editorial Office
MDPI AG
Grosspeteranlage 5
4052 Basel, Switzerland

This is a reprint of the Special Issue, published open access by the journal *Polymers* (ISSN 2073-4360), freely accessible at: https://www.mdpi.com/journal/polymers/special_issues/Advances_Textile_Structural_Composites_II.

For citation purposes, cite each article independently as indicated on the article page online and as indicated below:

Lastname, A.A.; Lastname, B.B. Article Title. *Journal Name* **Year**, *Volume Number*, Page Range.

ISBN 978-3-7258-4505-7 (Hbk)
ISBN 978-3-7258-4506-4 (PDF)
https://doi.org/10.3390/books978-3-7258-4506-4

Contents

About the Editor

Rajesh Kumar Mishra

Prof. Rajesh Kumar Mishra works at the Czech University of Life Sciences, Prague, Czech Republic. He has more than 26 years of academic and research experience in the polymeric material engineering field. His research areas are textile structural composites, green composites, nanocomposites, biomechanical engineering of polymeric structures, thermo-mechanical characterization of polymeric materials, etc.

polymers

MDPI

Editorial

Advances in Textile Structural Composites II

Rajesh Kumar Mishra

Department of Material Science and Manufacturing Technology, Faculty of Engineering, Czech University of Life Sciences Prague, 16500 Prague, Czech Republic; mishrar@tf.czu.cz

Textile structural composites are increasingly being recognized in engineering fields due to their exceptional mechanical attributes, lightweight nature, and improved durability when compared to conventional materials. These composites consist of textile structures from glass, carbon, aramid, or natural fibers, that are integrated within a polymer, metal, or ceramic matrix, leading to outstanding strength-to-weight ratios, resistance to corrosion, and versatility in design. The mechanical properties of such composites are influenced by factors such as the type of textile reinforcement, fiber orientation, and interfacial bonding, which affect characteristics like tensile strength, impact resistance, and fatigue performance [1]. These qualities are further enhanced by hybrid composites, which blend various fabric types to provide increased structural efficiency. The final composite characteristics and structural integrity are affected by a number of production processes, including pultrusion, vacuum-assisted resin infusion (VARI), resin transfer molding (RTM), manual lay-up, and additive manufacturing. By maximizing resource use and cutting waste, emerging fabrication techniques like 3D printing and automated fiber placement are completely changing the business. Textile-based composites are used in several engineering applications in the sports equipment, automotive, marine, aerospace, and civil infrastructure sectors. Their high-performance qualities help reduce weight and improve durability and energy efficiency [2].

They are utilized in the automobile industry for body panels, chassis components, and crash-resistant structures for enhanced vehicle performance and safety, and in aerospace for fuselage panels, wing structures, and interior components. Boat hulls and offshore constructions are examples of marine applications that take advantage of their resilience to severe weather conditions. To ensure lifespan and improved performance, civil engineering incorporates fiber-reinforced composites (FRCs) into bridge decks, structural rehabilitation, and seismic retrofitting [3]. The high strength-to-weight ratio, corrosion resistance, ease of fabrication, and improved design flexibility are just a few of the many benefits of FRCs. However, issues like moisture absorption, recycling difficulty, environmental effect, and high initial prices continue to stand in the way of wider implementation. To overcome these constraints, efforts are being made to improve material efficiency and environmental sustainability through the development of sustainable bio-based composites, sophisticated recycling methods, and computer modeling. FRCs' role in high-performance engineering applications will grow as a result of their ongoing development, which is being fueled by breakthroughs in material science and creative production techniques [4].

In order to guarantee their widespread integration into next-generation engineering solutions, this Special Issue included futuristic research directions and possible advancements while providing a thorough overview of textile structural composites, examining their performance, fabrication processes, applications, benefits, and drawbacks.

In addition to their structural and mechanical benefits, textile-reinforced composites contribute to the creation of smart and multifunctional materials, which integrate additional characteristics such as self-healing, impact sensing, and temperature management. These composites can be modified with nanoparticles, conductive fibers, and embedded sensors to improve damage detection and predictive maintenance in important structures [5]. Industries like defense and biomedical engineering are studying these composites for ballistic protection, prostheses, and bio-inspired applications where durability and adaptability are critical [6]. The future of FRCs seems bright as long as research into sustainable reinforcements, artificial intelligence-driven material design, and computational modeling continues [7]. To standardize testing procedures, increase their manufacturability, and guarantee their wider acceptance across engineering domains, increased cooperation between academic institutions, business, and regulatory agencies is crucial. Engineers can unleash the potential of next-generation materials and create innovations that redefine structural performance and material efficiency across industries by incorporating sustainability and digital advances into the creation of textile-reinforced composites [8].

This Special Issue is the subsequent addition to a previous Special Issue on the same topic and continues to provide further insights into advanced textile geometrical structures used as reinforcement in composites. In this Special Issue, the computational and geometrical models that can effectively predict the failure occurring in the textile reinforcement composite structures were discussed.

The applications and potential of statistical evaluation in textile-based composites were elaborated. The valorization and utilization of textile wastes for creating value-added composites was given utmost importance for sustainable product development. Bio-based materials and their importance for textile structures as well as composites were further discussed. Unconventional and modern application areas, e.g., biomedical, electronic gadgets, communication, sensors, etc., were explored.

The Editor is thankful to all the contributors and editorial staff for preparing this Special Issue successfully and effectively. These specialized contributions will lead to new methods of research and development in these emerging areas of sustainable composite product development.

Conflicts of Interest: The author declares no conflicts of interest.

References

1. Jamshaid, H.; Mishra, R.K.; Ahmad, N.; Nadeem, M.; Muller, M.; Kolar, V. Exploration of Effects of Graduated Compression Stocking Structures on Performance Properties Using Principal Component Analysis: A Promising Method for Simultaneous Optimization of Properties. *Polymers* **2022**, *14*, 2045. [CrossRef] [PubMed]
2. Riba, J.-R.; Cantero, R.; Riba-Mosoll, P.; Puig, R. Post-Consumer Textile Waste Classification through Near-Infrared Spectroscopy, Using an Advanced Deep Learning Approach. *Polymers* **2022**, *14*, 2475. [CrossRef] [PubMed]
3. Mittal, P.; Gokhale, S.T.; Manjunath, S.; Al-Qahtani, S.M.; Magbol, M.A.; Nagate, R.R.; Tikare, S.; Chaturvedi, S.; Agarwal, A.; Venkataram, V. Comparative Evaluation of Locally Administered 2% Gel Fabricated from Lemongrass Polymer and 10% Doxycycline Hyclate Gel as an Adjunct to Scaling and Root Planing in the Treatment of Chronic Periodontitis—A Randomized Controlled Trial. *Polymers* **2022**, *14*, 2766. [CrossRef]
4. Yong, S.; Hillier, N.; Beeby, S.P. Phase-Inverted Copolymer Membrane for the Enhancement of Textile Supercapacitors. *Polymers* **2022**, *14*, 3399. [CrossRef] [PubMed]
5. Hassan, T.; Jamshaid, H.; Mishra, R.; Khan, M.Q.; Petru, M.; Tichy, M.; Muller, M. Factors Affecting Acoustic Properties of Natural-Fiber-Based Materials and Composites: A Review. *Textiles* **2021**, *1*, 55–85. [CrossRef]
6. Uusi-Tarkka, E.-K.; Skrifvars, M.; Khalili, P.; Heräjärvi, H.; Kadi, N.; Haapala, A. Mechanical and Thermal Properties of Wood-Fiber-Based All-Cellulose Composites and Cellulose-Polypropylene Biocomposites. *Polymers* **2023**, *15*, 475. [CrossRef] [PubMed]

7. Mishra, S.; Chaturvedi, S.; Ali, M.; Pandey, K.K.; Alqahtani, N.M.; Alfarsi, M.A.; Addas, M.K.; Vaddamanu, S.K.; Al Ahmari, N.M.; Alqahtani, S.M.; et al. Dimensional Stability of Light-Activated Urethane Dimethacrylate Denture Base Resins. *Polymers* **2023**, *15*, 744. [CrossRef] [PubMed]
8. Mínguez-García, D.; Díaz-García, P.; Gisbert-Payá, J.; Bonet-Aracil, M. Emulsion Nanofibres as a Composite for a Textile Touch Sensor. *Polymers* **2023**, *15*, 3903. [CrossRef] [PubMed]

polymers

MDPI

Article

Exploration of Effects of Graduated Compression Stocking Structures on Performance Properties Using Principal Component Analysis: A Promising Method for Simultaneous Optimization of Properties

Hafsa Jamshaid [1], Rajesh Kumar Mishra [2,*], Naseer Ahmad [1], Muhammad Nadeem [1], Miroslav Muller [2] and Viktor Kolar [2]

1 Faculty of Textile Engineering, National Textile University, Faisalabad 37610, Pakistan; hafsa@ntu.edu.pk (H.J.); naseerahmad@ntu.edu.pk (N.A.); mnadeem6258@gmail.com (M.N.)
2 Department of Material Science and Manufacturing Technology, Faculty of Engineering, Czech University of Life Sciences Prague, Kamycka 129, 165 00 Prague, Czech Republic; muller@tf.czu.cz (M.M.); vkolar@tf.czu.cz (V.K.)
* Correspondence: mishrar@tf.czu.cz

Abstract: This paper focuses on the comfort properties of graduated and preventive compression stockings for people who work long hours in standing postures and for athletes for proper blood circulation. The present study was conducted in order to investigate the effects of the yarn insertion density and inlaid stitches on the performance of the compression stockings. The effects of these parameters on the thermo-physiological comfort properties were tested with standard and developed methods of testing. All compression stockings were maintained with class 1 pressure as per German standards. The structural parameters of the knitted fabric structures were investigated. The stretching and recovery properties were also investigated to determine the performance properties. The theoretical pressure was predicated using the Laplace's law by testing the stockings' tensile properties. The compression interface pressures of all stockings were also investigated using a medical stocking tester (MST) from Salzmann AG, St. Gallen, Switzerland. Correlation between the theoretical pressures and pressures measured using the MST system were also assessed. The current research used a multi-response optimization technique, i.e., principal component analysis (PCA), to identify the best structure based on the optimalization of the above-mentioned properties. The results also revealed that samples with higher insertion density levels exhibit better comfort properties. The results showed that sample R1 was the best sample, followed by R2 and P. In addition, all developed stocking samples exhibited better comfort properties than the control sample from the market.

Keywords: graduated compression stockings; medical textile; principal component analysis; thermo-phycological comfort; inlay yarn; insertion density

1. Introduction

The successful usage of compression garments has been in practice for a long time in order to prevent medical issues such as scarring, edema and deep vein thrombosis (DVT) in hospitalized patients, either immediately after surgery or because of immobility due to a medical illness [1,2]. DVT can be prevented with the use of compression stockings instead of drugs. Drugs can cause bleeding, which is a primary concern in patients who have undergone surgery. Compression can be applied using bandages and elastic compression hosieries (ECHs). Compared with bandages, ECHs are generally more in demand due to their convenience and ease of care. Compression hosiery products can be developed using circular and flat knitting machines. These are functional products that combine knitting technologies to meet healthcare requirements. Graduated compression stockings (GCS) help prevent the formation of blood clots in the legs by applying varying amounts of pressure to different parts of the leg [3,4]. Compression stockings are known as an effective nonsurgical choice, exerting gradual pressure to prevent and to treat venous disease in legs [5]. In the field of healthcare, compression stockings play a significant role, being used as a first line of treatment to manage the increasing intensity of varicose veins or chronic venous diseases at the initial stages [4–6]. Varicose veins are abnormal, whereby dilated blood vessels develop due to weaknesses in the vessel walls. This appears as swollen, twisted clusters or clots of blue or purple veins, which cause the blood to back up and pool inside the vein and make a new route [7]. Compression stockings exert an external pressure on lower limbs, reduce the vein diameter and increase the venous blood flow velocity, which helps the vein valves to function properly [8]. Researchers concluded that for compression stockings, $\geq$20 mmHg is an effective range of optimum pressure values that cause the fastest venous flow in various regions. There are different values for the ankle, calf, knee, lower thigh and upper thigh regions, which are 18 mmHg, 14 mmHg, 10 mmHg and 8 mmHg, respectively [9].

Compression stockings can be worn for an entire day because they maintain compressive pressure and do not lose their elastic stretch recovery over many hours. Changes in body posture significantly influence the skin pressure profile [10]. Nowadays, the biggest challenge for compression products and new devices is to make the compression more comfortable and acceptable for patients [11]. Usually, comfort is related to three aspects, namely thermo-physiological, sensorial and psychological factors. The sweat absorption and drying abilities are the main components of thermo-physiological comfort. It has been observed that the structural parameters of the knitted fabrics play an important role in determining the moisture management and thermal properties [12,13]. Researchers investigated the effects of the materials and structural parameters of compression stockings on the skin pressure distribution via material testing and wear trials [14]. Bruniaux et al. developed different models to study the effect of yarn on compression [15].

Cieślak et al. analyzed the impacts of pressure measurement methods along with two types of knitting structures, i.e., warp knitting and weft knitting, on the compression values using the same materials and concluded that that pressure values were lower in the warp-knitted fabric as compared to the single-jersey weft-knitted fabric in both vivo and vitro tests [16]. Troynikov et al. studied the effects of the physical attributes of warp-knitted tricot and single-jersey weft-knitted products on compression values. They concluded that the degree of compression pressure directly influences the construction and fit of the garment, the structure and physical properties of the material and the size and shape of the garment [17]. Liu et al. studied plain and rib structures developed on circular sock machines and conducted in vitro and in vivo testing of the compression pressure, finding consistent results [18]. Partsch et al. studied the compression and stiffness parameters of different compression stockings that involved European standards classes 1, 2 and 3 with pressure ranges of 15–21, 23–32, 34–46 mmHg, respectively [19]. Alisauskiene et al. investigated the effects of the linear density and insertion density of double-covered inlay yarn on the mechanical behavior (such as the washing and drying shrinkage percentage and compression behavior) of weft-knitted orthopedic supports [20]. Sarı et al. investigated the effects of process parameters, i.e., elastane yarn fineness, feeding tension and tightness factor, on the pressure of compression stockings [21]. Similarly, in another study researchers investigated the effects of the elastane yarn count on the extensibility and bursting strength of stockings [22].

The development of compression stockings is a complex process, and most of the scientific knowledge available is primarily focused on the study of readymade stockings, pressure evaluations and material properties [4,14,19,23,24]. Although compression stockings have gained popularity in recent years, the knowledge available on their performance in terms of thermo-physiological comfort remains scarce.

The aim of the present study was to develop compression stockings with different patterns and structures by changing the inlay yarn density and laid-in stitches while keeping the compression class constant. Compression testing was carried out using two different methods and their correlation was established. Stretch and recovery% values were also investigated for all samples. Additionally, the effects of the inlay yarn density and laid-in stitches on thermo-physiological comfort properties were also studied. All of the developed samples were compared with the control sample from the market. Furthermore, the current study focused on the optimization of the above fabric properties using principal component analysis (PCA). The study is expected to provide solutions for more comfortable compression stockings for patients.

2. Materials and Methods

2.1. Materials

Compression stocking samples were purchased from different commercial brands, e.g., TED® (D) and Sockey® (C) of Germany and China, respectively. The purchased samples were of compression class I, which is more commonly referred to as the standard in clinical application [25]. Class I is used for minor varicose veins, venous insufficiency and mild edema. All of the samples available on the market have a plain–single-jersey (1 × 1) structure (without insertion). Such a sample was used as a control for comparison purposes. The yarn specifications of the developed samples are shown in Table 1. The analysis was performed with great precision and accuracy under standard atmospheric conditions.

Table 1. Yarn specifications.

Sample	Yarn Parameters	Units	Plaiting Yarn (SCV) Linear Density (Denier)	Inlay Yarn DCV Linear Density (Denier)	Main Yarn/Ground Linear Density (Denier)
Developed samples	Overall Linear density	Denier	92 ± 2	320 ± 2	150 ± 2
	Top Covering linear density	Denier	70 ± 1	20 ± 1	-
	Bottom covering linear density	Denier	-	20	-
	Elastane (Lycra) linear density	Denier	20 ± 1	160 ± 2	150 ± 2
	Type	-	Multifilament	Multifilament	Multifilament
	Nylon No. of filaments	-	24	20*2	48
	Composition PU/PA	Fraction	0.08/0.92	0.56/0.44	0.0/1.0
	Color	-	Raw white	Raw white	Raw white
	Cross section	-	Round	Round	Round

Due to the different machine settings, V-shaped socks were produced using three types of yarn, i.e., main, plaiting and inlay yarns. In this research work, three types of fiber materials were used, i.e., 100% polyamide (PA 6.6) filament yarn as the main yarn, polyurethane (PU(Lycra®))/PA6.6 single-covered (SCV) yarn as a plaiting yarn or ground yarn and PU(Lycra®)/PA6.6 double-covered (DCV) yarn as an inlay yarn, as mentioned in Figure 1. All yarns were imported from China.

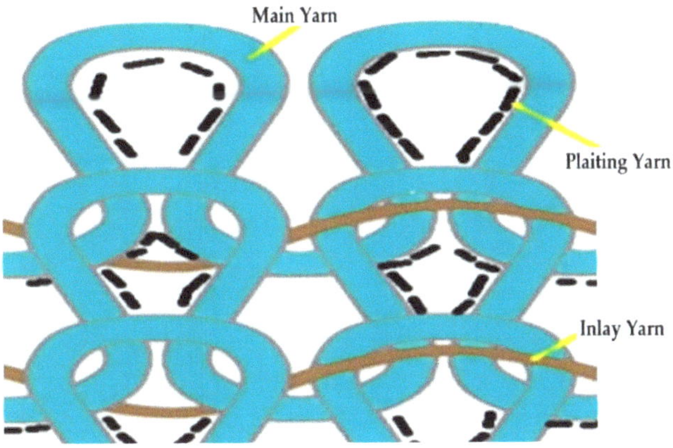

Figure 1. Schematic diagram of yarn.

The yarn specifications are given in Table 1.

2.2. Methods

Compression stockings were manufactured using a professional circular compression knitting machine. The machine was a Merz CC4-2 instrument from Germany, with a 12 cm diameter, E 24 machine gauge, 360 needles and 4 feeders, including two furnishers. The first feeder has 10 fingers, while the second, third and fourth feeders have 8 fingers. For the development of compression stockings, inlay yarn was fed at the first and third feeders, while the plaiting yarn and main yarn were fed through the second and fourth feeders, respectively. At feeder no. 1, in finger 2 and feeder 8, nylon filaments yarn was used for foot formation. At feeder no. 3, double-covered yarn in finger no. 4 and nylon filament yarn in finger 6 were fed for top welt formation. The inlay yarn arrangement is the main decisive factor in the manufacturing of different knitted structures. Compression stockings were made in plain and mock rib structures. A total of six samples were developed, representing the plain, rib 2 × 2 and rib 1 × 3 structures. The samples differed according to their inlay yarn insertion pattern (laid in stitch) and structure. In the first 3 structures, with sample codes P, R1 and R2, the inlay yarn was inserted after one empty course, while in other three samples with codes IP1, IR1 and IR2, the inlay yarn was inserted in each course. At the start the welt was formed and after 2–3 steps or courses the welt was completed. The graduated pressure regions denoted as points C, B1 and B in Figure 2 were manufactured as per the program. In the final step, the foot portion was manufactured.

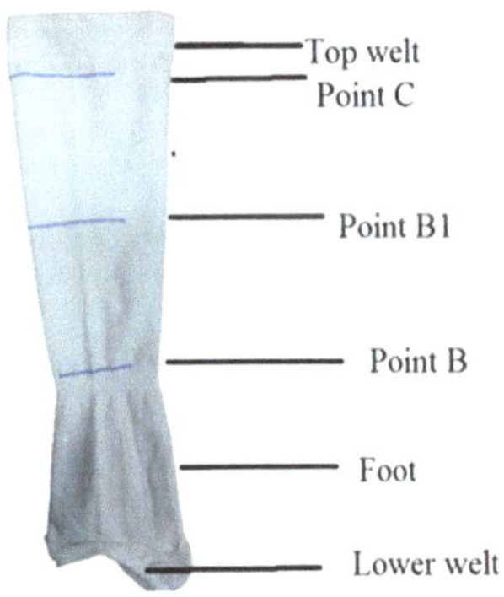

Figure 2. Graduated compression stocking with different compression levels.

The compression stockings were made in plain and mock rib structures (longer inlay threads floating at the back, forming a mock rib visual effect). The samples and their codes are listed in Table 2.

Table 2. Experimental samples and codes.

Sample No.	Sample Code	Sample Name/ID
1	P	Plain
2	R1	Rib 2 × 2
3	R2	Rib 3 × 1
4	IP1	insertion every course Plain
5	IR1	Insertion Rib 2 × 2
6	IR2	Insertion Rib 3 × 1

Loop configurations and structures and fabric simulations were constructed using the SDS-ONE APEX platform (Shima Seiki, Wakayama, Japan), as shown in Table 3 [26]. The corresponding microscopic images from the front and back views were also taken using an OPTIKA C-B 10 microscope (via Rigla, 30 24010 Ponteranica BG-Italy, SN 536495) with an M-144 fixed microscope adaptor.

Table 3. Structural representation and systematic diagrams of the developed samples.

Sample Code	Sample Id	Loop Structure	Fabric View		Microscopic View	
			Front	Back	Front	Back
P	Plain	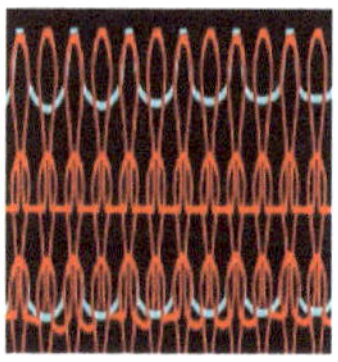		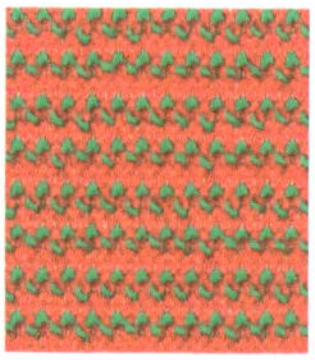		
R1	Rib 2 × 2	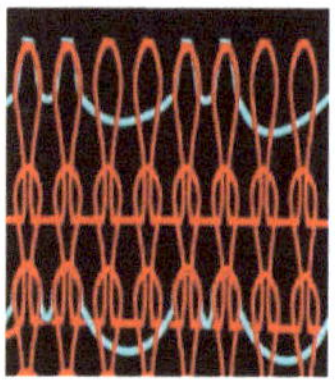				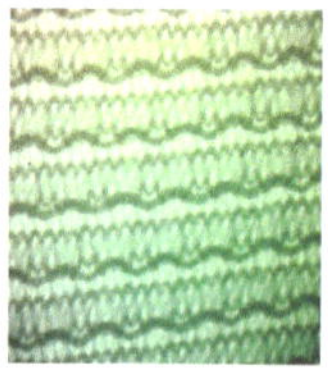
R2	Rib 1 × 3	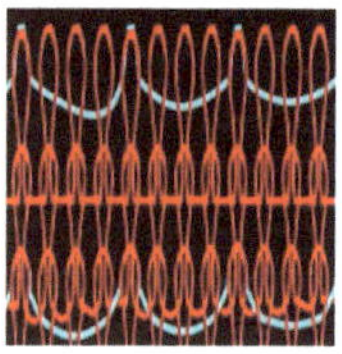		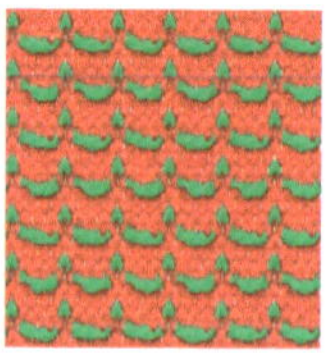		

Table 3. *Cont.*

Sample Code	Sample Id	Loop Structure	Fabric View		Microscopic View	
			Front	Back	Front	Back
IP1	Insertion Plain 1 × 1	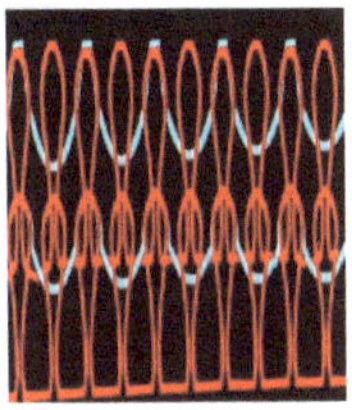		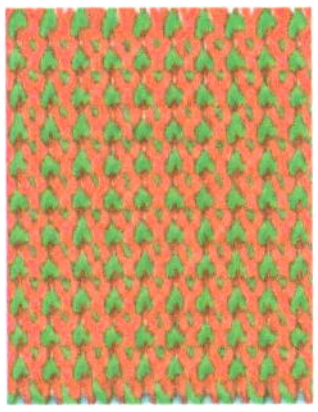		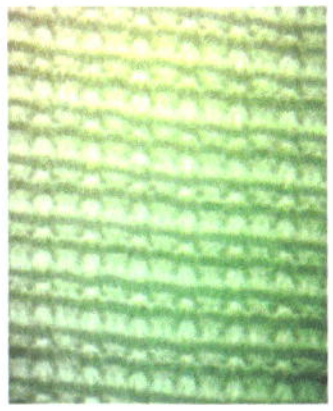
IR1	Insertion Rib 2 × 2	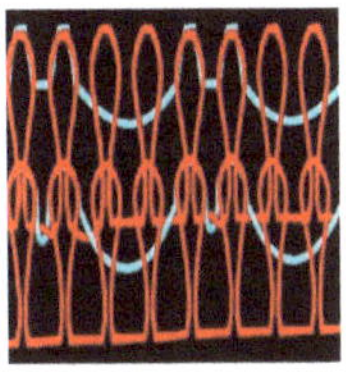				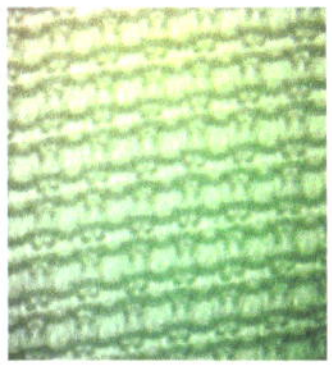
IR2	Insertion Rib 1 × 3	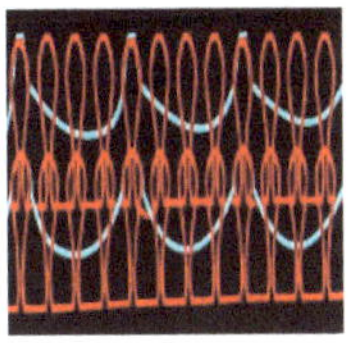		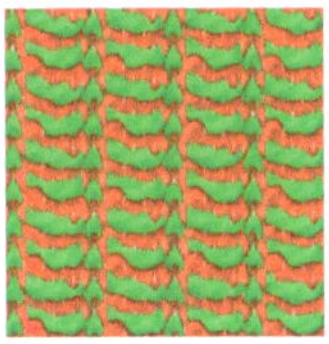		

All of the developed samples were of compression class 1 (C1), i.e., 18–21 mmHg and medium (M) size. The graduated pressure percentages for the class 1 samples were in line with Quality Assurance RALGZ 387/1 [27], as shown in Table 4.

Table 4. Gradients of the studied compression stockings [18,27].

Sr. No	Pressure Points	Required Pressure in Percentage
1	B (ankle)	100
2	B1	70 to 100
3	C	50 to 80
5	D, E, F, G	20 to 60

In the present study, the developed calf (below-knee) stockings contained 3 pressure regions, i.e., B, B1 and C. The pressure levels are shown in Figure 3.

Washing Method

The samples were washed according to Australian guidelines AS-2001:5-2005. Mild detergent was used in water and a soapy solution was prepared. The washing agent should not contain any bleaching agent, additives or chlorine because such substances weaken the elastic behavior of compression socks [28]. The samples were soaked in mild detergent solution for 5 min. Next, the stockings were shaken in the liquid for 2–2.5 min and then rinsed off with water. The excessive water was removed from the stockings by pressing them between the palms gently and then they were dried for 24 h on a flat surface. The washing and drying process was performed at room temperature. All testing was performed after washing.

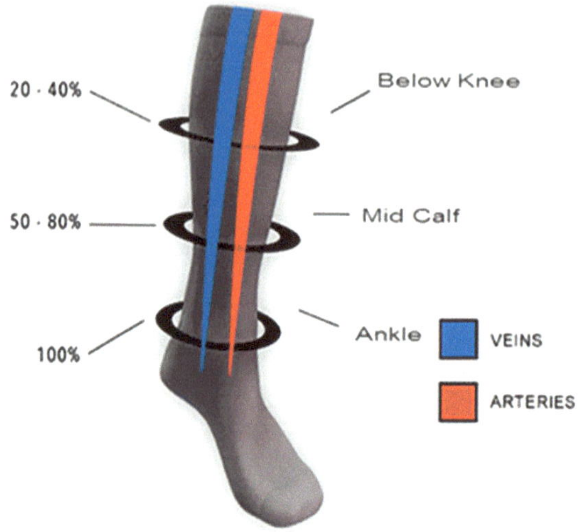

Figure 3. Pressure levels in the compression stocking.

2.3. Testing

2.3.1. Yarn Linear Density

Linear density values of the yarns were tested as per ASTM D 1059-17, and the numbers of filaments were counted using a microscope (Beck London 35288, Model 47, London, UK).

2.3.2. Physical Properties

Testing was performed after washing and at all reference points, as shown in Figure 2. These explanations are given only for point B, as this is the main reference point of interface pressure and also shows the maximal extension during movement. An analysis was performed of the physical properties of the stockings, i.e., courses per centimeter (CPC), wales per centimeter (WPC), areal density (GSM) and thickness. Courses and wales per cm were counted using counting glass as per laboratory practice. Thickness values were measured as per ASTM D1777, using a 99-0697 thickness tester (Framincham, MA, USA). To determine the areal density values (g/m^2) of stockings, a model JH-10-36 GSM cutter (Jenhaur Co. Ltd., Taipei, Taiwan) was used to cut samples according to ASTM D 3776.

2.3.3. Stretch and Recovery Percentages

The stretch and recovery ability is very important for properly fitting socks. Stretching is also directly linked to the comfort of a garment, as it relates to the ease of wearing a garment. The optimum level of stretchiness is a basic requirement in knitted stockings to support blood flow in the legs and feet. This is one of the most important performance characteristics. Poor stretching results in a poor fit, leading to performance shortcomings such as discomfort. Compression stockings should be designed such that they maintain a uniform interface pressure gradient over the limb for effective recovery results [29,30]. Stretch measurements of all compression stockings were performed on a CETME Attrezzature Per Calzific instrument from Reggio Em., Italy, as per ASTM D 2594. The stretchiness is the amount of extension of the fabric under a determined force, while the elasticity is the fabric's recovery ability after stretching, which determines the dimensional stability of an elastic garment.

Stretch and recovery % values were calculated using Equations (1) and (2).

The stretch percentages were calculated using Equation (1):

$$\text{Fabricstretch\%} = (B - A)/A \times 100 \tag{1}$$

where A is the original distance between marked points prior to the application of tension, B is the distance between benchmark points on the specimen under tension.

For the calculation of the recovery percentages, Equation (2) was used:

$$\text{Fabricrecovery\%} = (B - D)/(B - A) \times 100 \tag{2}$$

where A is the original distance between marked points prior to the application of tension, B is the distance between benchmarks on the specimen under tension and D is the distance between benchmarks after the release of tension.

2.3.4. Shrinkage %

All the samples were tested in circumference and transversal directions after washing.

2.3.5. Compression Testing

Compression testing can be performed in two ways in vivo (direct or wearer method) and in vitro (indirect or model method) using different instruments, such as a HOSY system, tensile testing device, HATRA and medical stocking tester (MST). In the present research, an in vitro method was adopted to evaluate the compression pressure values. A Sazlmann AG stocking measuring device available in different MST models (e.g., MST, MST II, MST III, MST IV and MST V models) (Salzmann AG, St Gallen, Switzerland, MST MK V), as shown in Figure 4, was used in the present research. The measurement principle of MST is based on the use of a pressure transducer. The compression testing was performed according to 387/1-2008, using quality and testing specifications for medical compression hosiery prescribed by RAL German Institute for Quality Assurance and Labeling [27]. A pneumatic probe or sensor was placed between a wooden leg and stocking samples. Compression pressure values were measured at the pressure points, e.g., the ankle (B), gaiter (B1) and calf (C). The samples were placed vertically along a wooden leg model with a circular cross-section. In compression garments, mmHg is commonly used as the unit of compression pressure to define the pressure dosage (1 mmHg = 133.322 Pascal). All pressure values were noted in mmHg. The pressure values were measured when each reading was stable. Three readings were taken at each point and the average was used for analysis. Three measurements were taken at each measurement point after an interval of three minutes.

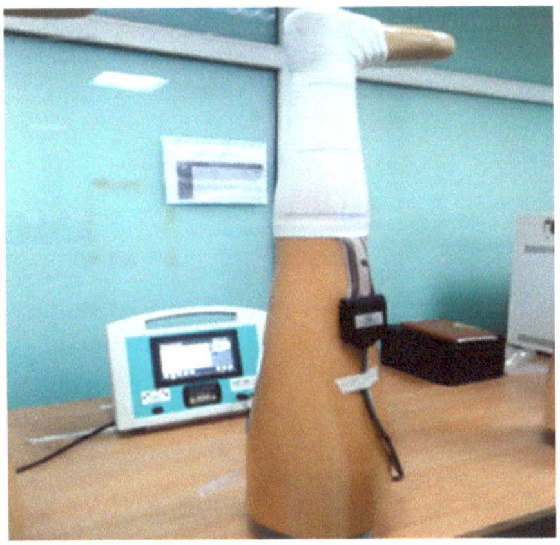

Figure 4. A medical stocking tester (MST) and MK V pressure measurement device.

All compression stockings were designed on the principle of elastic compression. Compression stockings on the leg generate the pressure with which the stocking presses the leg. Laplace's law explains how the pressure is exerted by stockings on the limb. It has been reported that pressure exerted by a strip on the surface of a human leg can be determined by Laplace's law. This law is now widely used to explain and assess the pressure delivered to a limb of a known radius by a fabric under known tension [31,32]. Equation (3) is shown below:

$$P = T/r \tag{3}$$

where P denotes pressure (Pa), T is the tension of the compression material on the leg (Nm^{-1}), r is the radius (m) of the limb surface on which it is applied and W is the width in meters. Mathematical models were developed on the basis of Laplace's law. Compression measurements were also carried out using a tensile testing device to compare the two methods. A tensile testing machine (LLOYD, LRX Plus, Largo, FL, USA) was used for measurements of fabric tension during extension [21]. Extension in the course direction was chosen because compression stockings are stretched in the transversal direction to generate pressure during wear. The tensile properties were evaluated at pressure point "B" using the same machine settings. The device gauge length was set to 86 mm, while extension using a constant traverse speed of 100 mm/min and load cell of 100 N was used. The specimen width was 8 cm and the radius was 4 ± 0.4 cm for all samples. A comparison was made between calculated pressure and experimental measurements using an MST instrument.

2.3.6. Thermo-Physiological Comfort Properties

The air permeability (AP) of all samples was tested using an air permeability tester (M021A, SDL Atlas, Newark, NJ, USA) as per the ISO-9237 standard under a pressure of 100 Pa. In total, ten readings were taken and the average was calculated.

Moisture management of the samples was checked using a moisture management tester (M 290 SDL Atlas, Newark, NJ, USA) as per the AATCC-195 standard. In total, three readings were taken and their average was calculated.

Thermal properties of samples were checked by Permetest, Liberec, Czech Republic as per standard ISO-11092. Five readings were taken, and average was calculated.

All experiments were carried out at a standard atmosphere, i.e., 21 ± 1 °C temperature and 65 ± 2% relative humidity in compliance with ISO 139:2005 standard.

2.3.7. Data Analysis

Means and standard deviations were computed for all test results. In the process of manufacturing a standardized product, variability is considered the main hurdle in achieving the desired product quality. Different researchers have proven the process optimization technique as the best solution for not only improving the quality but also reducing error in the development of the final product. The optimization techniques include grey relational analysis (GRA), principal component analysis (PCA), desirability function, artificial neural network (ANN), fuzzy logic, data development and analysis and weighted signal-to-noise (WSN) approaches [33,34]. PCA explains the variation in a set of parameters through a few linear combinations called principal components in order to reduce the original data set and refine the linear relations that exists among the variables. Each principal component explains the percentage of overall variation among the variables. These principal components are ranked in descending order, from the maximum to the minimum. In PCA, multiple response variables are converted using eigenvalues as weights into a single response, called a multi-response performance index (MRPI), which is computed by adopting the following procedure. MRPI values are computed by first computing the signal-to-noise and normalized signal-to-noise ratios of each response using the standard method described in earlier studies [35].

The samples and their codes are shown in Table 2. The finishing experiments were conducted in three independent replicates for each sample and the data reported are the mean values of three readings. In the present study, PCA was employed to identify the best parameters for samples subject to the simultaneous optimization of multiple responses in the development of stockings, including compression by MST, air permeability, thermal resistance, overall moisture management capacity (OMMC), stretch %, recovery %, shrinkage %, thickness and areal density.

3. Results and Discussion

The physical features, dimensional properties (i.e., shrinkage, stretch and recovery %) and thermo-physiological comfort properties of the control sample from the market (denoted as C) and of the developed samples (P, R1, R2, IP1, IR1, IR2) were analyzed.

In total, 6 sets of experimental runs and corresponding experiments on the nine responses were performed, including compression by MST (mmHg), air permeability (AP in mm/sec), thermal resistance (TR in m^2K/W), OMMC, stretch (%), recovery (%), shrinkage (%), thickness (mm) and areal density (g/m^2), with the results given in Table 5.

Table 5. Details of experiments and mean experimental results.

Experimental Run	Sample	Responses								
	Sample Code	Y1: Compression by MST (mmHg)	Y2: AP (mm/s)	Y3: TR (m^2K/W)	Y4: OMMC	Y5: Stretch (%)	Y6: Recovery (%)	Y7: Shrinkage (%)	Y8: Thickness (mm)	Y9: Areal Density (g/m^2)
1	P	19.4	534.06	0.03	0.58	63.73	98.55	−2.5	0.75	221
2	R1	21.1	572.07	0.02	0.59	65.03	98.55	−1.31	0.94	220
3	R2	18.7	689	0.01	0.79	62.5	100	0	0.99	239
4	IP1	24.5	488	0.02	0.64	62.86	94.69	−2.5	0.71	257
5	IR1	20.03	504.03	0.02	0.71	63.42	100	0	0.9	260
6	IR2	22	627.05	0.04	0.77	62.5	96	−6.25	0.87	252

The fabric areal density and thickness values of all samples (P, R1, R2, IP1, IR1, IR2 and C) are compared in Figure 5.

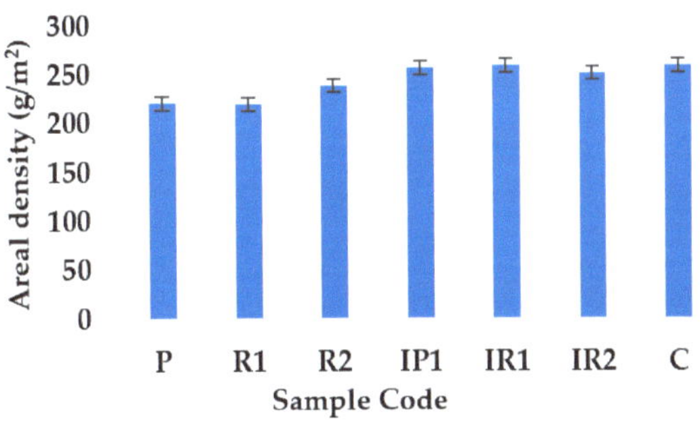

(a) Areal density

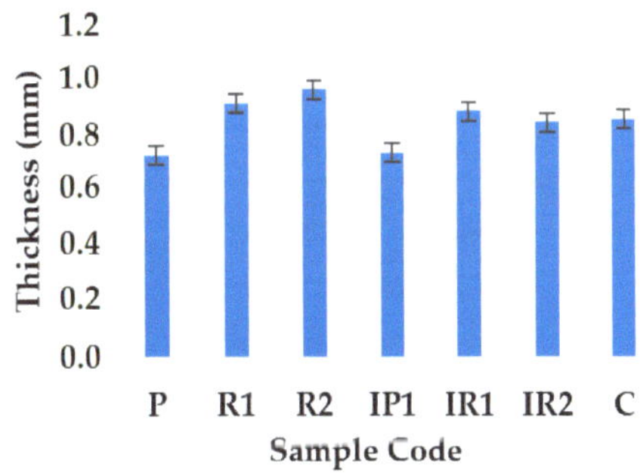

(b)Thickness

Figure 5. Physical properties of graduated stocking at point B.

The areal density and thickness of the fabric are important properties. The areal density and thickness were measured and compared on the basis of the pattern of insertion of the inlay yarn and structure, as shown in Figure 5. Samples IP1, IR1 and IR2 were found to have higher areal density or GSM values due to the insertion of inlay yarn in every course, which led to increased fabric weight per unit surface area. As the number of tuck stitches increases, the accumulation of yarn in the tucking zones increases, which leads to increases in weight. In 1 × 1 samples, the inlay yarn was tucked in one wale and floated in the next wale consecutively. In 2 × 2 samples, the inlay yarn was tucked into two wales followed by two floating loops. In 1 × 3 samples, the inlay yarn was tucked in one wale followed by three floating loops. The effect of the thickness on the pattern was not significant. Patterns with lower insertion density values were found to have higher levels of thickness. This may be attributed to the increased bulkiness of the samples. By increasing the number of missing stitches in the inlay yarn, the thickness was increased, as was observed by other researchers [36]. The properties of the developed samples were compared with the control sample (C) from the market.

3.1. Stretch % and Recovery %

As the inlay yarn insertion density increased, the stretch % and recovery % decreased. This may have been due to the tight structures, whereby movement of yarn was restricted. This trend is illustrated in Figure 6. The stretching is inversely proportional to the tightness. As the structure becomes tighter, the level of stretching decreases. In the 1 × 1 structure, 1 floating loop length alternatively appeared in the wale, while in the other wales it was tucked. A tucked wale is less stretchable than a floating wale. The reason for the lower level of stretching was the alternative tuck pattern of the inlay yarn in the wale direction. In the 2 × 2 structure, the two consecutive floating loops of inlay yarns led to greater stretching in the fabric. These two floats do not pass through any loops when the force is applied. Thus, the fabric stretches without any restriction from the knitted loops. In the 1 × 3 structure, there were three consecutive floating loops of inlay yarn, which were then tucked in the same wale. This arrangement increases the restriction to movement and reduces the extensibility [36].

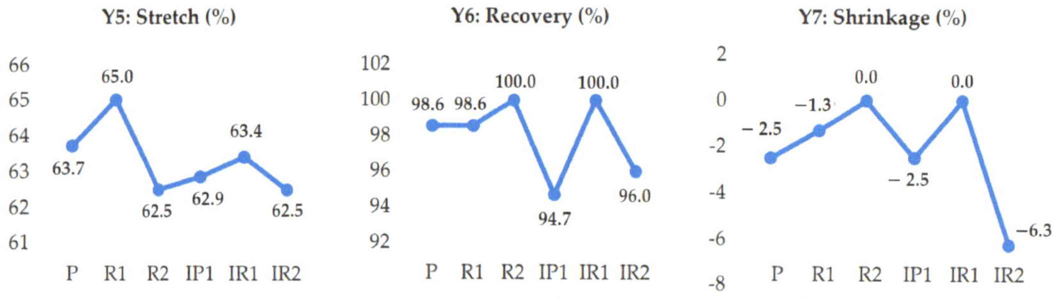

Figure 6. Dimensional properties of compression stockings.

3.2. Shrinkage %

Shrinkage is also an important factor influencing the dimensional stability during washing. Due to the elastomeric yarns, the fabric dimensions changed. The circumferences of the stockings were reduced after washing due to shrinkage in the knitted structures. The shrinkage is reduced if a structure is knitted with a smaller number of inlay yarns. It can be observed that with an increase in inlay yarn, the circumference of the stocking is reduce. Shrinkage of less than +6% is not considered significant. Normally, increasing the tuck loop results in higher stability and shape retention in knitted structures [37]. The recovery % values of all tested samples were found to be above +94%, which implies that the compression textiles sustained their properties after washing.

3.3. Compression Measurement

The compression pressure measurements were performed using two methods. All samples satisfied the requirements for class 1 compression, irrespective of the inlay yarn insertion density and structure.

Using the MST method, compression testing was performed on each pressure point in the control and developed samples. Using the tensile testing method, compression was measured only at point B in the developed samples. Point B is the most crucial pressure point for compression stockings. As we move from point B to B1 and B1 to C, the structure becomes wider as the stitch length increases; thus, the compression value decreases. This phenomenon is known as gradual compression. During knitting and testing, it was verified that the compression of all the samples remained gradual and within the class 1 pressure range according to the RAL 387/1 standard. This pressure range helps the blood to flow in the upward direction due to the lower pressure at higher points.

From the results, it is clear that a knitted pattern where inlay yarn is inserted in every course has better compression properties. Samples where the inlay yarn is inserted in alternating courses generate lower levels of compression, as supported by other researchers [36,37]. Xiong et al. [38] also stated that the inlay yarn provides the required level of compression in elastic fabrics, while the main yarn provides the stiffness and thickness. To achieve a higher level of compression, it is necessary to increase the density of the inlay yarn. Tables 6 and 7 summarize the results.

Table 6. Average pressure values.

Sample Code	Measuring Points	Compression Values (mmHg)
C	B	20.1
	B1	14.2
	C	13.5
T	B	16.1
	B1	14.9
	C	12.5
P	B	19.4
	B1	14.3
	C	13.2
R1	B	21.1
	B1	15.4
	C	11.1
R2	B	18.7
	B1	12.5
	C	10.7
IP1	B	24.5
	B1	16.8
	C	11.2
IR1	B	20
	B1	17
	C	12
IR2	B	22
	B1	17.3
	C	13.7

Note: The pressure exerted on the leg: 1 mmHg = 133.3 Pa.

Table 7. Comparison of results using both methods.

Sample Code	By MST MKV	By Laplace Equation (mmHg)	Difference	Deviation %
P	19.4	19.66	−0.26	−1.34021
R1	21.1	20.95	0.15	0.7109
R2	18.7	17.22	1.48	7.914439
IP	24.5	30.02	−5.52	−22.5306
IR1	20	31.52	−11.52	−57.6
IR2	22	28.08	−6.08	−27.6364

The pressure was calculated using Laplace's equation and was measured using MST (R = 4.07 cm) compression at point B [39,40].

The deviation percent was calculated as the difference between the measured compression using the MST and pressure calculated using Laplace's equation. The results in Table 6 show that the deviation values when applying Laplace's equation for point B ranged from ±0.7% to ±57.6%. There was significant deviation when applying Laplace's equation in all samples when compared with the practically measured compression values using the MST method. The deviation was higher in points where the inlay yarn insertion was increased. This is because more pressure is exerted with a higher inlay yarn insertion density. The theoretical pressure values calculated using the equation did not accurately predict the values found using the experimental method due to several limitations, as explained by other researchers [41–44]. The correlation coefficient of the interface pressure using both experimental and theoretical methods was found to be 0.632, as shown in Figure 7. Lower R values depict smaller correlations between the two methods.

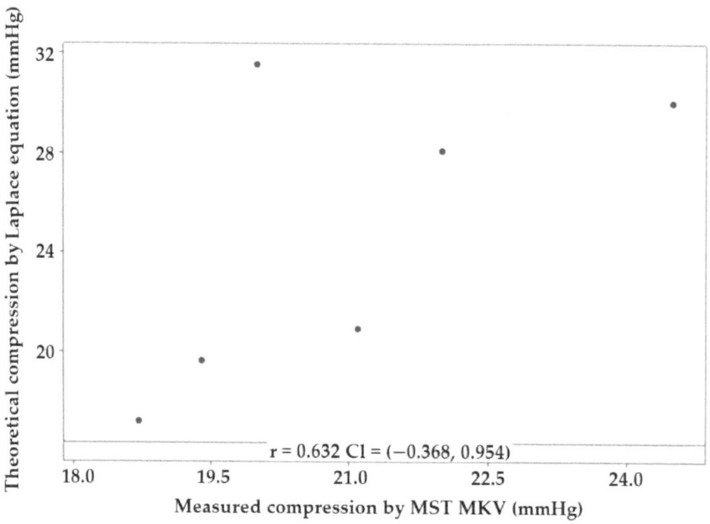

Figure 7. Correlation between measured and calculated compression pressure.

3.4. Comfort Properties

3.4.1. Air Permeability

The air permeability is an important property with respect to comfort, especially regarding the breathability of compression stockings. These stockings are worn for the majority of the day, meaning a lack of comfort negatively affects performance. Under real conditions during wear, stretching of the stocking would cause increased interloop spaces, which would also affect air permeability. From point B to B1 and from B1 to C, the structure becomes wider the and stitch length increases. As a result, the air permeability increases.

From the Figure 8a, it is clear that stocking samples without insertion show higher air permeability due to the lack of insertion of the inlay yarn, as compared to samples 4–6, where insertion of the inlay yarn is performed in each course. The inter-yarn pores are the most important components, which influence the porosity of the fabric structure and facilitate the passage of air.

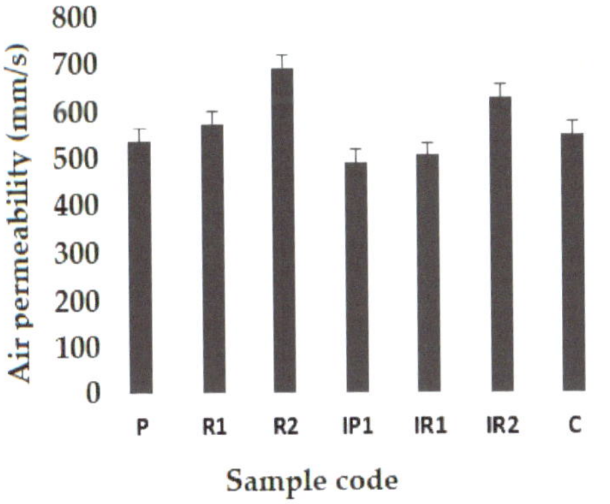

(a) Air Permeability

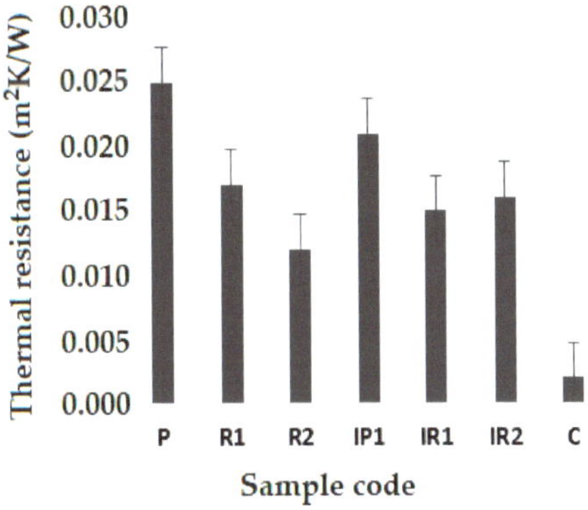

(b) Thermal Resistance

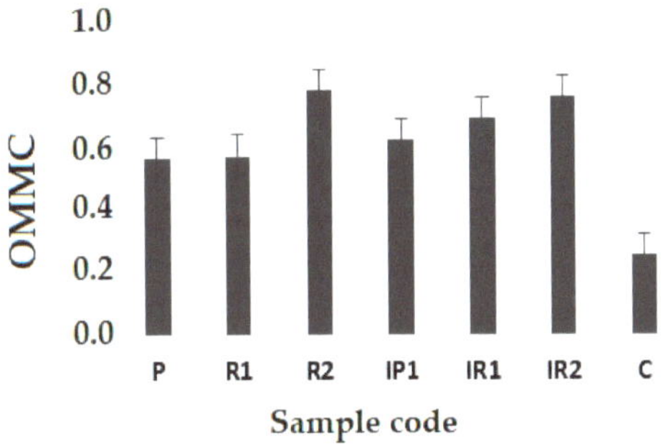

(c) Overall Moisture Management Capacity (OMMC)

Figure 8. Thermo-phycological comfort properties of the developed stockings.

The type of structure also has a strong influence on the air permeability, as it affects the spaces in the loop, which determine the porosity and ultimately the air permeability. This can be seen in the optical or microscopic images of fabrics shown in Table 3.

Due to the arrangement of tucked and missed stitches in the inlay yarn in the developed knitted structures, a visible difference in air permeability can be observed. In the 1 × 1 structure, the inlay yarn is tucked in the wales alternatively with a floating loop. In the 2 × 2 rib structure, the inlay yarn is fed in two wales followed by two floating loops. In the 1 × 3 rib structure, the inlay yarn is fed into one wale followed by three floating loops. Due to the longer float length of the inlay yarn, spaces in the loops and the pore size increase, which leads to higher air permeability. Missed stitches in knitted fabrics result in lower surface area and allow the air to pass through easily. The commercial control sample (C) had a 1 × 1 structure without insertion and had comparable air permeability to the developed samples.

3.4.2. Thermal Resistance

While moving from point B to C, the structure becomes wider as the tightness decreases and stitch length increases. This results in emptier spaces in the structure at higher points, while the thermal resistance decreases.

It can be observed from Figure 8b that samples without the insertion of inlay yarn in every course show higher thermal resistance than samples with insertion. The reason may be due to the lower thickness of theses samples. The fabric thickness has a profound effect on the thermal resistance. The type of structure has a significant effect on the thermal resistance. More compactness in the structure results in higher thermal resistance, as stagnant air is entrapped in small pores. Stagnant air has the lowest thermal conductivity when compared with other fibers and solids. It has been previously reported that dead air pockets in knitted structures increase the thermal resistance of the fabric [45].

The thermal resistance of stagnant air is higher than that of fiber. From the microscopic images, it is evident that sample R2 has bigger pores. Therefore, air cannot be entrapped in it due to the longer floating loops as compared to R1 and P. The accumulation of yarn due to tucked stitches affects the dead air pockets, which increases the thermal resistance. As the float length increases, the dead air pockets decrease. The commercial sample had lower thermal resistance than the developed samples.

3.4.3. Overall Moisture Management Capacity (OMMC)

Appropriate moisture management is essential for stockings to prevent the formation of wet conditions, which leads to reduced tolerance of the human tissues to shear stress and friction. A higher moisture content and sweating lead to uncomfortable conditions for the wearer. Moisture management is a very important property facilitating the evaporation of generated sweat. Overall, the developed stockings exhibited reasonably good moisture management properties. The samples showed OMMC values higher than 0.5, which were in the good (0.4–0.6) and very good grade range (0.6–0.8).

The commercial sample had an OMMC value lower than 0.5, making it uncomfortable to wear in summer season.

From the Figure 8c, it can be observed that the samples with the insertion of inlay yarn in every course exhibit lower OMMC values than samples without insertion. This is due hindrance of the vapor transfer due to the presence of tuck loops. These loops may stop the flow of water in capillary spaces. The 1×1 structure shows a lower OMMC value than the other structures, e.g., 2×2 and 1×3. Due to the alternating tuck and miss stitches, the transmission of vapor decreases, which leads to an increased wetting time and decrease in spreading speed. With more tuck points, the 2×2 structure exhibits a higher OMMC value than the 1×1 knitted structure. This is due to the higher number of tuck points, which act as capillary channels for moisture transmission from inner to outer surfaces. As the density of the missed stitches increases, the OMMC also improves.

3.4.4. Principal Component Analysis (PCA) of the Results

The multi-response optimization of the nine responses was computed by performing the following steps.

In the first step, the signal-to-noise ratio of each response in each run was commuted depending on the quality characteristics by adopting the standard method. The results are shown in Table 8. For comparison purposes, the normalized signal-to-noise ratios of all responses were computed depending on the quality characteristic, e.g., the higher the better, the lower the better or the more nominal the better. In the present study, the quality characteristics are compression MST (mmHg), AP (mm/s), OMMC, stretching (%) and recovery (%). For these characteristics, the responses were taken as the higher the better. However, for thermal resistance (TR in m^2K/W), shrinkage (%), thickness (mm) and arial density (g/m^2), the responses were set as the lower the better. Normalized signal-to-noise ratios are given in Table 9.

Table 8. Signal-to-noise (S/N) ratios of the responses.

Factor		Responses								
Run	Sample Code	Compression (mmHg)	AP (mm/s)	TR (m^2K/W)	OMMC	Stretch (%)	Recovery (%)	Shrinkage (%)	Thickness (mm)	Areal Density (g/m^2)
1	P	25.76	54.55	0.03	−4.79	36.09	39.87	−7.96	2.54	−46.89
2	R1	26.49	55.15	0.02	−4.64	36.26	39.87	−2.33	0.57	−46.85
3	R2	25.44	56.76	0.01	−2.05	35.92	40.00	−1.58	0.12	−47.57
4	IP1	27.78	53.77	0.02	−3.93	35.97	39.53	−7.96	3.01	−48.20
5	IR1	26.03	54.05	0.02	−3.02	36.04	40.00	−1.58	0.92	−48.30
6	IR2	26.85	55.95	0.04	−2.24	35.92	39.65	−15.92	1.24	−48.03

Table 9. Normalized S/N ratio.

Factor		Responses								
Run	Sample Code	Compression (mmHg)	AP (mm/s)	TR (m^2K/W)	OMMC	Stretch (%)	Recovery (%)	Shrinkage (%)	Thickness (mm)	Areal Density (g/m^2)
1	P	0.136	0.261	0.546	0.000	0.492	0.732	0.445	0.165	0.027
2	R1	0.447	0.461	0.257	0.055	1.000	0.732	0.052	0.844	0.000
3	R2	0.000	1.000	0.000	1.000	0.000	0.993	0.000	1.000	0.496
4	IP1	1.000	0.000	0.415	0.315	0.145	0.000	0.445	0.000	0.931
5	IR1	0.254	0.094	0.164	0.646	0.369	1.000	0.000	0.724	1.000
6	IR2	0.602	0.727	1.000	0.933	0.001	0.252	1.000	0.611	0.813

As discussed earlier, PCA explains the variation in a set of parameters via linear combinations. Fewer variables are preferred in order to reduce the original linear relations between the variables. Each principal component (PC) corresponds to a percentage of the total variation. In this study, PC1 represents the linear combination that explains the maximum variance (43.6%), followed by PC2, PC3 and PC4 (52.1%). The last five PCs (PC5–PC9) contribute to less than 5% of the variance. The first four PCs represent almost 96% of the total variance among the data set, as shown in Table 10. The next step is to decide the number of PCs retained.

Table 10. Eigenvalues of the six remaining objectives.

	PC1	PC2	PC3	PC4	PC5	PC6
Eigenvalue	3.9269	2.7173	1.4085	0.5704	0.377	0
Proportion	0.436	0.302	0.156	0.063	0.042	0
Cumulative	0.436	0.738	0.895	0.958	1	1

As per the Kaiser–Guttman principle, the first three principal components (PC1, PC2, PC3) with eigenvalues greater than one were kept for further analysis (Table 10).

The normalized signal-to-noise ratios were taken for the analysis using Minitab 17. The first three components had eigenvalues greater than one, as shown in Table 10. The scree plot shown in Figure 9a was taken for further analysis. The normalized S/N ratios were used to perform the multivariate analysis using Minitab 17. The first and second principal components had eigenvalues greater than one. Thus, they were considered for further analysis. The principal component analysis matrix is given in Table 11. The results are presented in Figure 9b.

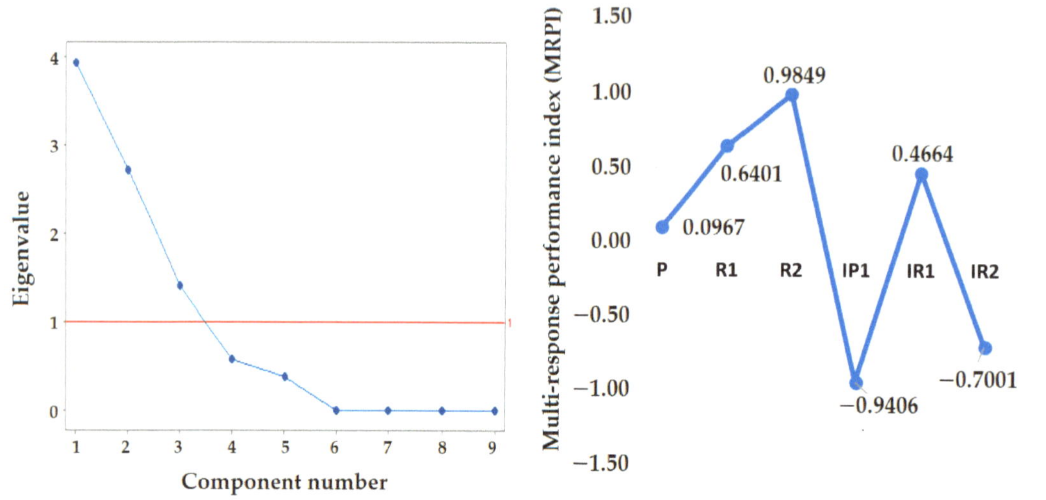

(a) Scree plot (b) Multi-response performance index (MRPI)

Figure 9. Scree plot and MRPI values.

Table 11. Principal component analysis matrix.

Principal Component	Eigen Value	Proportion	Eigen Vector								
First	3.9269	0.436	0.425	0.306	−0.385	−0.635	0.482	0.476	−0.413	0.383	−0.225
Second	2.7173	0.302	0.1	−0.407	−0.079	−0.598	0.5	−0.021	−0.16	−0.274	−0.331
Third	1.4085	0.156	−0.193	0.451	0.472	−0.108	0.164	−0.031	0.412	0.1	−0.563

From the results given in Table 11, the order of the multi-response performance index (MRPI) values for the second, third and first runs was obtained as first, second and third highest, respectively, as illustrated in Table 12 and Figure 9b. This implies that samples R1, R2 and P may be more effective than samples IP1, IR1 and IR2.

Table 12. Normalized S/N ratios and MRPI values.

Factor		Responses										
Run	Sample Code	Compression (mmHg)	AP (mm/s)	TR (m^2K/W)	OMMc	Stretch (%)	Recovery (%)	Shrinkage (%)	Thickness (mm)	Areal Density (g/m^2)	MRPI Values	Order
1	P	0.136	0.261	0.546	0.000	0.492	0.732	0.445	0.165	0.027	0.386	3
2	R1	0.447	0.461	0.257	0.055	1.000	0.732	0.052	0.844	0.000	1.330	1
3	R2	0.000	1.000	0.000	1.000	0.000	0.993	0.000	1.000	0.496	0.415	2
4	IP1	1.000	0.000	0.415	0.315	0.145	0.000	0.445	0.000	0.931	−0.258	5
5	IR1	0.254	0.094	0.164	0.646	0.369	1.000	0.000	0.724	1.000	0.369	4
6	IR2	0.602	0.727	1.000	0.933	0.001	0.252	1.000	0.611	0.813	−0.741	6

4. Conclusions

The present study investigated the effects of the inlay yarn insertion density levels in different knitted structures used as compression stockings. The compression values and their comfort-related properties were evaluated at the graduated compression points. For healthcare and medical usages, the application of inlay yarn and laid-in stitches is crucial in order to provide the desired pressure in preventive stockings. From the results, it was concluded that the developed samples had better comfort-related properties than the commercial sample taken from the market. The pressure was evaluated theoretically using Laplace's law and estimated experimentally using a medical stocking tester (MST). The correlation was established with an R value of 0.632.

This study utilized a principal component analysis (PCA) for the simultaneous optimization of properties in compression stockings. The simultaneous optimization process suggested that stockings with inlay yarn insertion in alternate courses will show optimum comfort results. The best sample exhibited a maximum air permeability of 700 mm/s, minimum thermal resistance of 0.013 m^2K/W and overall moisture management capacity (OMMC) of 0.8, showing far superior results to the control sample collected from the market. In this study, a total of 9 principal components were used. PC1 represented the linear combination that explains the maximum variance (43.6%), followed by PC2, PC3 and PC4 (52.1%). The last five PCs (PC5–PC9) contributed to less than 5% of the variance, while the first four PCs represented more than 95% of the total variance among the data set. As per the Kaiser–Guttman principle, the first three principal components (PC1, PC2, PC3) with eigenvalues greater than one were kept for further analysis.

Consequently, when designing preventive graduated stocking with class 1 compression, it is necessary to choose the most appropriate inlay yarn insertion density and structure combination. This approach is essential in order to improve the wearable comfort properties and to achieve economic benefits by using the minimum amount of inlay yarn, as it contains elastane and is expensive. In the future, more research is necessary to further improve the comfort and dimensional stability of compression stockings by changing the stitch lengths and linear density levels of the different yarns.

Author Contributions: H.J., R.K.M., N.A., M.N., M.M. and V.K., experiment design; H.J., R.K.M., N.A., M.N., M.M. and V.K., testing of mechanical properties and data analysis; H.J., R.K.M., N.A., M.N., M.M. and V.K., writing and editing of the paper; H.J., R.K.M., N.A., M.N., M.M. and V.K., language correction; H.J., R.K.M., N.A., M.N., M.M. and V.K., revision of the paper; R.K.M., communication with the editors; H.J., R.K.M., N.A. and M.M., administration and funding; H.J., R.K.M., N.A., M.M. and V.K., supervision. All authors have read and agreed to the published version of the manuscript.

Funding: This work was supported by HEC Project TDF03-149, "Development of comfortable compression stockings for varicose veins or chronic venous disease" and the Internal grant agency of the Faculty of Engineering, Czech University of Life Sciences, Prague, No. 2022:31140/1312/3105: "Research on composite materials with polymer matrix and natural filler in the field of additive technology".

Institutional Review Board Statement: Not applicable.

Informed Consent Statement: Not applicable.

Data Availability Statement: Not applicable.

Conflicts of Interest: The authors declare no conflict of interest.

References

1. William, E.C., Jr. Preventing deep vein thrombosis in hospital inpatients. *BMJ Clin. Rev.* **2007**, *335*, 147–151. [CrossRef]
2. Lindsay, S.S.; Boghossian, R.S.; Stewart, M. Compression stockings for the initial treatment of varicose veins in patients without venous ulceration. *Cochrane Data System. Rev.* **2011**, *11*, 1–29.
3. Brijesh, N. Compression therapy for venous leg ulcers. *Indian Dermatol. Online J.* **2014**, *5*, 378–382. [CrossRef]
4. Guo, X.; Cui, H.; Wen, X. Treatment of varicose veins of lower extremity: A literature review. *Int. J. Clin. Exp. Med.* **2019**, *12*, 2142–2150.
5. Siddique, H.F.; Mazari, A.A.; Havelka, A.; Kus, Z. Performance Characterization and Pressure Prediction of Compression Socks. *Fiber Polym.* **2020**, *21*, 657–670. [CrossRef]
6. Weiss, R.A.; Duffy, D. Clinical benefits of lightweight compression: Reduction of venous-related symptoms by ready-to-wear lightweight gradient compression hosiery. *Dermatol. Surg.* **1999**, *25*, 701–704. [CrossRef]
7. Available online: https://my.clevelandclinic.org/health/diseases/4722-varicose-veins (accessed on 3 April 2021).
8. Available online: https://www.healthline.com/health/compression-stockings-for-varicose-veins#how-compression-stockings-workn (accessed on 3 April 2021).

9. Lawrence, D.; Kakkar, V. Graduated, static, external compression of the lower limb: A physiological assessment. *Br. J. Surg.* **1980**, *67*, 119–121. [CrossRef]
10. Bochmann, R.P.; Seibel, W.; Haase, E. External compression increases forearm perfusion. *J. Appl. Physiol.* **2005**, *99*, 2337–2344. [CrossRef]
11. Mervis, J.S.; Lev-Tov, H. *Compression Therapy, in Local Wound Care for Dermatologists*; Springer: Berlin/Heidelberg, Germany, 2020; pp. 83–94.
12. Ahmad, H.S.; Jamshaid, H. Development of thermo-physiologically Comfortable Knit Structure for Sports Application. *Tekst. Ve Konfek.* **2019**, *29*, 105–112. [CrossRef]
13. Ghosh, A.; Mal, P.; Majumdar, A.; Banerjee, D. Analysis of Knitting Process Variables and Yarn Count Influencing the Thermo-physiological Comfort Properties of Single Jersey and Rib Fabrics. *J. Inst. Eng. India Ser. E* **2016**, *97*, 89–98. [CrossRef]
14. Liu, R.; Kwok, Y.L.; Li, Y.; Lao, T.T. Skin pressure profiles and variations with body postural changes beneath medical elastic compression stockings. *Int. J. Dermatol.* **2007**, *46*, 514–523. [CrossRef]
15. Bruniaux, P.; Crepin, D.; Lun, B. Modeling the mechanics of a medical compression stocking through its components behavior: Part 1—Modeling at the yarn scale. *Text. Res. J.* **2012**, *82*, 1833–1845. [CrossRef]
16. Cieślak, M.; Karaszewska, A.; Gromadzińska, E.; Jasińska, I.; Kamińska, I. Comparison of methods for measurement of the pressure exerted by knitted fabrics. *Text. Res. J.* **2017**, *87*, 2117–2126. [CrossRef]
17. Troynikov, O.; Wardiningsih, W.; Watson, C.; Luca, O. Influence of material properties and garment composition on pressure generated by sport compression garments. *Procedia Eng.* **2013**, *60*, 157–162. [CrossRef]
18. Liu, R.; La1, T.T.; Wang, S. Technical Knitting and Ergonomical Design of 3D Seamless Compression Hosiery and Pressure Performances in Vivo and in Vitro. *Fiber. Polym.* **2013**, *14*, 1391–1399. [CrossRef]
19. Partsch, H.; Partsch, B.; Braun, W. Interface pressure and stiffness of ready made compression stockings: Comparison of in vivo and in vitro measurements. *J. Vasc. Surg.* **2006**, *44*, 809–814. [CrossRef] [PubMed]
20. Alisauskiene, D.; Mikucioniene, D. Investigation on alteration of compression of knitted orthopaedic supports during exploitation. *Mater. Sci.* **2012**, *18*, 362–366. [CrossRef]
21. Sarı, B.; Oğlakcıoğlu, N. Analysis of the parameters affecting pressure characteristics of medical stockings. *J. Ind. Text.* **2018**, *47*, 1083–1096. [CrossRef]
22. Ozbayrak, N.; Kavuşturan, Y. The effects of inlay yarn amount and yarn count on extensibility and bursting strength of compression stockings. *Text. Appl.* **2009**, *19*, 102–107. Available online: https://dergipark.org.tr/en/pub/tekstilvekonfeksiyon/issue/23634/251724 (accessed on 24 April 2022).
23. Stolk, R.; Wegen, C.P.M.; Franken, V.; Neumann, H. A method for measuring the dynamic behavior of medical compression hosiery during walking. *Dermatol. Surg.* **2004**, *30*, 729–736. [CrossRef]
24. Liu, R.; Kwok, Y.L.; Li, Y.; Lao, T.T.; Zhang, X. Quantitative assessment of relationship between pressure performances and material mechanical properties of medical graduated compression stockings. *J. Appl. Polym. Sci.* **2007**, *104*, 601–610. [CrossRef]
25. Lim, C.S.; Davies, A.H. Graduated compression stockings. *Can. Med. Assoc. J.* **2014**, *186*, E391–E398. [CrossRef] [PubMed]
26. Available online: https://www.shimaseiki.com/product/design/virtual_sampling/ (accessed on 19 April 2022).
27. Medical Compression Hosiery, Quality Assurance RALGZ 387/1. Available online: http://www.tagungsmanagement.org/icc/images/stories/PDF/ral_gz_387_englisch.pdf (accessed on 24 April 2022).
28. Available online: https://comprogear.com/washing-compression-socks (accessed on 15 April 2022).
29. Basra, S.A.; Rajput, A.W.; Zahid, B.; Asfand, N.; Azam, Z.; Jamshaid, H. Investigations of stretch and recovery properties of knitted socks. *Ind. Text.* **2020**, *71*, 132–137. [CrossRef]
30. Mishra, R.; Bhagti, V.; Behera, B. Novelties of 3-D woven composites and nanocomposites. *J. Text. Inst.* **2014**, *105*, 84–92. [CrossRef]
31. Schuren, J.; Mohr, K. The efficacy of Laplace's equation in calculating bandage pressure in venous leg ulcers. *Wounds UK* **2008**, *4*, 38–47.
32. Abdelhamid, R.R.; Brigita, K.S.; Zuhaib, A. Influence of Tensile Stress On Woven Compression Bandage Structure And Porosity. *AUTEX Res. J.* **2020**, *20*, 263–273. [CrossRef]
33. Ahmad, N.; Kamal, S.; Raza, Z.A.; Zeshan, M.; Abid, S.; Javed, Z.; Karahan, M. Comparative Analysis of the Performances of Six Taguchi-Based Multi-Response Optimisation Techniques for Product Development in Textiles. *Fibre. Text. East. Eur.* **2021**, *5*, 100–108. [CrossRef]
34. Selvakumar, S.; Arulshri, K.; Padmanaban, K.; Sasikumar, K. Design and optimization of machining fixture layout using ANN and DOE. *Int. J. Adv. Manuf. Technol.* **2013**, *65*, 1573–1586. [CrossRef]
35. Umair, M.; Shaker, K.; Ahmad, N.; Hussain, M.; Jabbar, M.; Nawab, Y. Simultaneous optimization of woven fabric properties using principal component analysis. *J. Nat. Fibers* **2017**, *14*, 846–857. [CrossRef]
36. Yu, A.; Sukigara, S.; Takeuchi, S. Effect of inlaid elastic yarns and inlay pattern on physical properties and compression behaviour of weft-knitted spacer fabric. *J. Ind. Text.* **2020**, 1528083720947740. [CrossRef]
37. Repon, M.R.; Laureckiene, G.; Mikucioniene, D. The Influence of Electro-Conductive Compression Knits Wearing Conditions on Heating Characteristics. *Materials* **2021**, *14*, 6780. [CrossRef] [PubMed]
38. Xiong, Y.; Tao, X. Compression Garments for Medical Therapy and Sports. *Polymers* **2018**, *10*, 663. [CrossRef] [PubMed]
39. Mikučionienė, D.; Milašiūtė, L. Influence of knitted orthopaedic support construction on compression generated by the support. *J. Ind. Text.* **2017**, *47*, 551–566. [CrossRef]

40. Rabe, E.; Partsch, H.; Junger, M. Guidelines for clinical studies with compression devices in patients with venous disorders of lower limb. *Eur. J. Vasc. Endovasc. Surg.* **2008**, *35*, 494–500. [CrossRef]
41. Chi, Y.W. A new compression pressure measuring device. *Veins Lymph.* **2017**, *6*, 6636. [CrossRef]
42. Mishra, R.; Jamshaid, H.; Yosfani, S.; Hussain, U. Thermo physiological comfort of single jersey knitted fabric derivatives. *Fash. Text.* **2021**, *8*, 40. [CrossRef]
43. Yao, B.; Li, Y.; Hu, J.; Kwok, Y.; Yeung, K. An Improved Test Method for Characterizing the Dynamic Liquid Moisture Transfer in Porous Polymeric Materials. *Polym. Test.* **2006**, *25*, 677–689. [CrossRef]
44. Raza, Z.A.; Ahmad, N.; Kamal, S. Multi-response optimization of rhamnolipid production using grey rational analysis in Taguchi method. *Biotechnol. Rep.* **2014**, *3*, 86–94. [CrossRef]
45. Cliff, N. The eigenvalues-greater-than-one rule and the reliability of components. *Psychol. Bull.* **1988**, *103*, 276–279. [CrossRef]

polymers

MDPI

Article

Post-Consumer Textile Waste Classification through Near-Infrared Spectroscopy, Using an Advanced Deep Learning Approach

Jordi-Roger Riba [1,*], Rosa Cantero [2], Pol Riba-Mosoll [1] and Rita Puig [2]

[1] Electrical Engineering Department, Universitat Politècnica de Catalunya, Rambla Sant Nebridi 22, 08222 Terrassa, Spain; pol.riba2@gmail.com
[2] Department of Computer Science and Industrial Engineering, Universitat de Lleida, Pla de la Massa 8, 08700 Igualada, Spain; rosa.cantero@udl.cat (R.C.); rita.puig@udl.cat (R.P.)
* Correspondence: jordi.riba-ruiz@upc.edu; Tel.: +34-937398365

Abstract: The textile industry is generating great environmental concerns due to the exponential growth of textile products' consumption (fast fashion) and production. The textile value chain today operates as a linear system (textile products are produced, used, and discarded), thus putting pressure on resources and creating negative environmental impacts. A new textile economy based on the principles of circular economy is needed for a more sustainable textile industry. To help meet this challenge, an efficient collection, classification, and recycling system needs to be implemented at the end-of-life stage of textile products, so as to obtain high-quality recycled materials able to be reused in high-value products. This paper contributes to the classification of post-consumer textile waste by proposing an automatic classification method able to be trained to separate higher-quality textile fiber flows. Our proposal is the use of near-infrared (NIR) spectroscopy combined with a mathematical treatment of the spectra by convolutional neural networks (CNNs) to classify and separate 100% pure samples and binary mixtures of the most common textile fibers. CNN is applied for the first time to the classification of textile samples. A total of 370 textile samples were studied—50% used for calibration and 50% for prediction purposes. The results obtained are very promising (100% correct classification for pure fibers and 90–100% for binary mixtures), showing that the proposed methodology is very powerful, able to be trained for the specific separation of flows, and compatible with the automation of the system at an industrial scale.

Keywords: textile waste; recycling; reuse; classification; deep learning; NIR spectroscopy; circular economy; post-consumer waste

1. Introduction

The textile industry is fundamental for society and everyday wellbeing (i.e., comfort and protection). Different textile products are needed, such as household products and personal clothes. Textile consumption has been exponentially growing in the last decade (having doubled its production around the world in the last 15 years) due to low costs and fast fashion [1,2]. Thus, large amounts of resources are extracted to produce clothes that will last less than one year (due to being disposed of or underutilized). Therefore, there is an urgent need for the implementation of new, sustainable textile consumption patterns (e.g., textile renting), although this may not be easy [3].

Production of textile products generates great environmental impacts along their value chain, including production of raw materials, manufacturing, and the products' end of life [1]. The main sources for raw materials in the textile industry are synthetic fibers (from oil, a non-renewable resource) and natural fibers (mostly from farming, such as of cotton, using great amounts of water and fertilizers) [4,5]. On the other hand, chemicals used during manufacturing processes (e.g., dyeing) [6] have the potential to cause environmental and human health impacts [7].

At their end of life, textile products can be reused, remanufactured, recycled, incinerated (with energy recovery), or landfilled. The degree of processing and the benefit for the environment vary from one option to the other [8]. Although material recovery should be prioritized [9], most of the textile clothes collected today are incinerated or landfilled, while a small part is reused as second-hand clothes when possible [10]. Material recovery needs a good classification method to separate the different materials for high-quality recycling.

Textile recycling rates (including reuse and remanufacture) account for only 13% of world clothing, and usually this small recycled amount is used in lower-value applications [1]. Textile industry manufacturing activities have experienced few changes in relation to traditional production systems, this being especially true in emerging countries, which account for the majority of production [11]. Recycling and reusing textiles could help to reduce the production of new textiles from raw materials and the associated use of chemicals, water, and energy in the production stage [12]. However, the shift toward a more circular textile industry still faces several challenges [13], especially in increasing the ratio of textile waste collection and recycling. Whereas textile consumption has considerably increased, textiles represent around 5% of landfill contents [14].

According to European legislation (Directive (EU) 2018/851) [15], selective collection of post-consumer textile waste should be implemented in all associated countries before January 2025, with the aim of increasing the recycling rate of textile waste in Europe. This represents a great challenge, due to the increasing complexity of textile composition [8].

Therefore, it is necessary to implement an automatic classification system to allow the correct separation of the different textile fibers, thus increasing their recyclability and the added value of recycled textile materials. This system will contribute to moving from the current linear textile system to a more circular one. Today, the classification of textiles is mainly performed manually (with high human resource costs) and rarely automated; thus, it is not possible to treat large amounts of waste each day [16].

The contributions of this paper are to the development of a sorting system able to separate the most common pure textile fibers (i.e., cotton, linen, wool, silk, polyester, polyamide, and viscose), as well as binary mixtures of viscose–polyester and cotton–polyester. The proposed sorting method is compatible with an automatic textile classification approach for textile fibers, and is based on a novel mathematical treatment (deep learning algorithms) of the near-infrared (NIR) spectra. Algorithms based on convolutional neural networks (CNNs) are applied to classify textile samples; NIR spectra can be registered through optical fiber sensors in a continuously working process, thus allowing the large-scale recycling of post-consumer textile waste. It should be noted that this approach can be applied for classifying and sorting many other materials, including plastics, paper, or leather, among others.

The near-infrared (NIR) region has a range from 750 to 2500 nm. The absorption of radiation is due to the combination of overtones and bands from the fundamental vibrations produced in the mid-infrared region [17]. These bands provide useful information, but have important limitations, such as their low intensity, which is about 2–3 times lower in magnitude compared to those in the mid-infrared. In addition, for a given molecule, different overtones and combination bands appear in the NIR region, producing a large number of overlapping bands, which makes it difficult to interpret.

Almost all of the absorption bands observed in NIR come from overtones of the stretching vibrations of AH_x groups or combination bands of these groups. They are mostly due to overtones of the CH, OH, and NH groups, although in some cases the bands corresponding to PH and SH groups can be observed.

The high degree of bands overlapping and the complexity of the spectrum, together with its dependence on the sample's physical state, make it difficult to perform simple calibrations (at one or a few wavelengths) when working with the NIR technique. Thus, the use of complex mathematical methods is needed to extract useful information from the spectrum. On the other hand, its dependence on chemical and physical parameters requires the calibration to be very complete, and to take into account all of the expected variability in the samples.

One of the great advantages of NIR spectroscopy is its versatility and adaptability to analyze samples of different natures. For solid samples, the measurement is carried out by diffuse reflectance. NIR radiation has the ability to significantly penetrate a solid sample and acquire important information. This is one of the characteristics that make this form of radiation an especially useful tool in the study of solids, as chemical information can be obtained in a non-invasive and non-destructive way.

When measurements are carried out by diffuse reflectance, the use of an optic-fiber probe allows the NIR spectrum to be recorded directly on the sample, without any previous treatment, achieving a remarkable saving of time and avoiding the use of polluting reagents.

NIR spectroscopy has experienced a strong advance in recent decades, motivated both by the development of new instruments—which allow the rapid recording of the full NIR spectrum—and by the implementation of mathematical methods for the processing of spectral information. Once calibrated, the NIR spectra can be used for the fast determination of the properties of the analyzed sample [18]. Thus, this technique is being used in very diverse industrial fields, such as food, biomedicine, pharmacy, petrochemicals, polymers, and the textile industry, among others [19–22].

Thus, the aim of this paper is to develop an industrially scalable sorting method to classify textile waste (composed of pure fibers and binary mixtures) by registering the NIR spectra of the samples, followed by a mathematical treatment using convolutional neural network (CNN) algorithms.

2. The Proposed NIR–CNN Approach and the Processing Methods

This paper analyzes the accuracy of two approaches for classifying unknown incoming fiber samples, based on a supervised strategy by analyzing the NIR spectral data (XDS™ Optiprobe Analyzer, FOSS AnalyticalA/S, Hillerod, Denmark) of such samples using appropriate mathematical methods. These two approaches are summarized in Figure 1.

According to Figure 1, the NIR spectrum of the unknown textile sample is preprocessed by applying the first or second derivative and mean-centering the obtained data. Next, two paths are analyzed: The first consists of directly applying a convolutional neural network (CNN) to the preprocessed spectral data to classify the sample into one of the preset classes. The second path is more complex since, after the preprocessing stage, the resulting spectral data are transformed through the principal component analysis (PCA) and canonical variate analysis (CVA) algorithms, to reduce their dimensionality and prepare the data for the classification stage, which is also achieved by means of a CNN.

2.1. Calibration and Prediction Data Subsets

The whole dataset is often split into calibration (or training) and prediction data subsets [23]. This strategy allows assessment of the accuracy of the applied classification methods by using a different set of samples (prediction subset) than the one used to calibrate or train the model (calibration subset). It is worth noting that the prediction subset is not used during the training stage. The whole set of available data was divided in a 50%/50% proportion, i.e., half of the samples were assigned to the calibration set, and the remaining half were assigned to the prediction set. This assignment was carried out at random.

For a fair training process and a good assessment of the results attained, the pertinence class of the whole set of samples must be known. Then, the samples belonging to the calibration subset must be labelled, with the labels assigning each sample to its class (i.e., type of textile fiber), since the proposed approach is a supervised training process. Although in this paper the pertinence class of the samples in the prediction subset is already known, because it is necessary to assess the performance of the classification process, in general, the pertinence class of the prediction samples is not known, and the classification method must accurately estimate their pertinence classes.

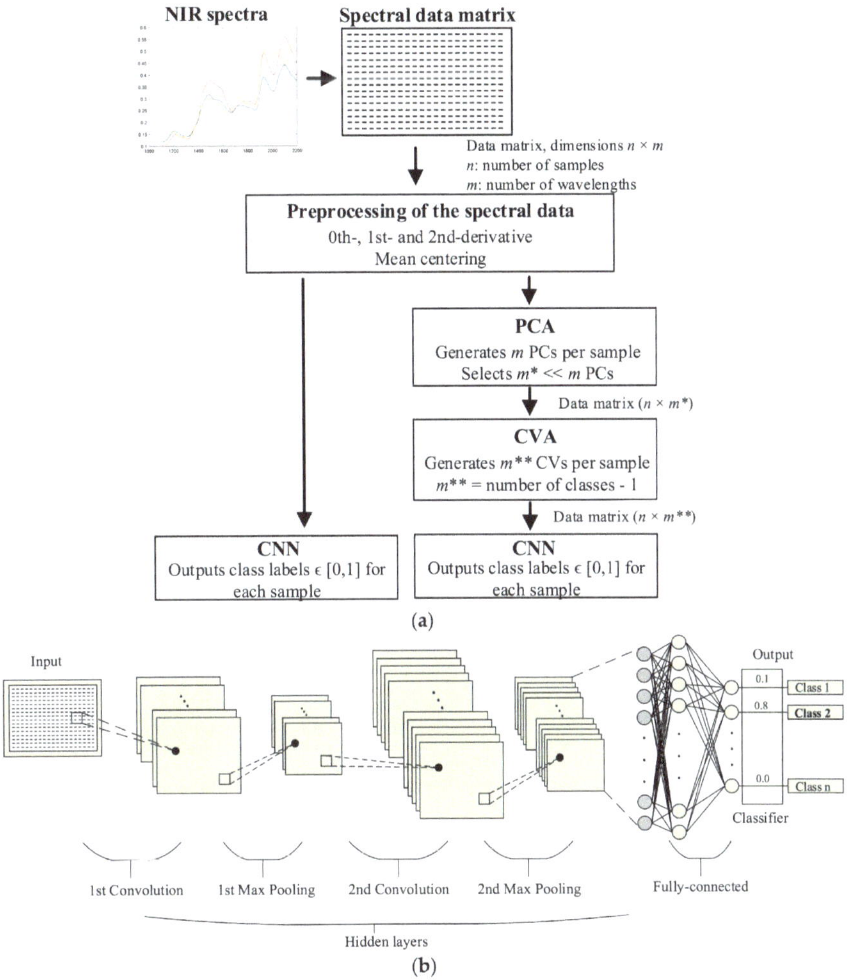

Figure 1. (**a**) Classification approach based on two different paths to classify an unknown fiber sample into one of the preset textile fiber classes from the NIR spectrum. (**b**) Structure of the classification dealt with by the CNN.

2.2. Data Processing Stage

It is a common practice to process the raw spectral data provided by the spectrometer [24] to increase the accuracy of the classification stage. It should be noted that spectral data are treated in matrix format. The matrix of raw data contains n rows (the number of samples) and m columns (the number of wavelengths in each spectrum).

The usual data processing consists of obtaining the first or second derivative of the spectra. Derivatives are applied with the aim of enhancing spectral differences. They act as a specific type of baseline correction, removing constant background signals at the cost of increasing noise. Derivatives are applied for enhancing the visual resolution, highlighting spectral structures, and resolving overlapping peaks. It is known that each derivative stage decreases the polynomial order by one. Whereas the first derivative allows the removal of a constant offset, the second derivative also removes a linear term of the spectrum [25]. The derivatives were obtained by means of the Savitzky–Golay smoothing and differentiation filter, which applies a moving average of 5 points to determine the first derivative and

10 points for the second derivative of the spectra. This means that the spectral data are analyzed in raw, first derivative, and second derivative modes.

Next, the columns of the data matrix are mean-centered to remove any bias, so that afterwards the mean-centered columns have zero mean, and their variances remain unaltered.

2.3. Applied Dimensionality Reduction Methods

Prior to the application of the CNN it is possible to apply dimensionality reduction methods, where dimensionality refers to the number of input variables or wavelengths in the analyzed problem. These methods aim at reducing the number of variables in the problem by combining the original variables into a condensed set of new variables. This step is important because NIR spectra include thousands of wavelengths per sample, and it is known that a large number of variables can reduce the performance of classification algorithms [26].

The new variables, also known as latent variables, are usually obtained by linearly combining the original variables, with the aim of retaining the same information and rejecting the inherent noise contained in the original variables.

In this paper, the multiclass canonical variate analysis (CVA) dimensionality reduction technique is applied. It condenses the significant spectral data into a small set of latent variables [27,28] calculated by linearly combing the original variables from the NIR spectrum. The basic idea of CVA is to calculate a set of projection vectors v optimizing the Fisher criterion—that is, to maximize the differences between different classes, while minimizing the intraclass variability [29]:

$$J(v_{\mathrm{CVA}}) = \max\left[\frac{v^T{}_{(1,m)} B_{(m,m)} v_{(m,1)}}{v^T{}_{(1,m)} W_{(m,m)} v_{(m,1)}}\right] \tag{1}$$

where v is a vector, m is the number of wavelengths in the NIR spectra, and B is the interclass dispersion matrix, defined as follows:

$$B_{(m,m)} = \sum_{i=1}^{c} n_i(\overline{x}_i - \overline{x})(\overline{x}_i - \overline{x})^T \tag{2}$$

and W is the intraclass dispersion matrix, defined as follows:

$$W_{(m,m)} = \sum_{i=1}^{c} \sum_{j=1}^{n_i} (x_{ij} - \overline{x}_i)(x_{ij} - \overline{x}_i)^T \tag{3}$$

where $\overline{x}_i = \sum_{i=1}^{c} x_{ij}/n_i$, with i = 1,2, … c, x_{ij} represents the elements of the spectral data matrix $X_{(n,m)}$, n is the number of samples, m is the number of original variables or wavelengths, c is the number of classes defined in the problem, and $\overline{x} = \sum_{i=1}^{c} n_i\overline{x}_i/n$, with $n = \sum_{i=1}^{c} n_i$.

From the matrix of projection vectors V, the new latent variables or canonical variates (CVs) can be determined as $Y_{(n,s)} = X_{(n,m)} V_{(m,s)}$, with $s = c - 1$.

One of the major drawbacks of CVA is the requirement that the number of samples be greater than the number of original variables. This requirement is often not fulfilled in problems based on spectral data, since NIR spectra include thousands of variables (wavelengths), while the number of samples is usually smaller. Therefore, it is imperative to apply a suitable algorithm for reducing the dimensionality of the problem, i.e., the number of variables before CVA can be applied. To this end, principal component analysis (PCA) is applied before applying the CVA algorithm, because PCA is probably the most recognized unsupervised algorithm for dimensionality reduction [30]. PCA allows calculation of a reduced number of uncorrelated and orthogonal latent variables, known as principal components (PCs) [31]. The PCs concentrate the significant information included in the original variables [32], while explaining most of the variance in the whole set of samples.

Although PCA calculates the same number of PCs as the number of original variables, only the first PCs explaining a predetermined amount of variance are retained [33], whereas the remaining PCs are not considered. To do so, the PCs must first be ranked in descending order, i.e., the first PC is that explaining the greatest variance, whereas the last PC is the one explaining the least variance. PCA is based on the singular value decomposition of matrix $X_{(n,m)}$:

$$svd(X) = U_{(n,n)} \Sigma_{(n,m)} V^T{}_{(m,m)} \quad (4)$$

and the new latent variables can be obtained as follows:

$$Y_{(n,m)} = X_{(n,m)} V_{(m,m)} \quad (5)$$

By applying the PCA + CVA sequence, the dimensionality of the problem is greatly reduced, to obtain a number of latent variables (CVs) equal to the number of classes minus one. Once the CVs are calculated, the CNN algorithm can be applied to classify the unknown textile samples, although the calculation of the PCA + CVA sequence is optional.

2.4. Convolutional Neural Networks (CNNs) for Classification

Artificial neural networks have shown their potential for successfully predicting the structural properties of textiles [34]. CNNs are regularized, fully connected neural networks. They were first applied to recognize handwritten numbers [35]. CNNs are widely applied in image classification, financial time series, natural language processing, or to classify waste from images [36], among many other applications. In this paper, CNNs are applied to classify textile samples in different classes according to their composition. The information only flows in one direction—i.e., from the input to the output—and the number of parameters that the CNN learns is fewer than in other deep learning networks, thus reducing the computational effort.

Figure 1b shows a basic diagram of a CNN, where the inputs can be videos, images, spectra, or any other signal type, and the outputs are the probabilities of the corresponding inputs belonging to each predefined class in the problem.

CNNs usually have three main parts, i.e., an input layer, a hidden layer, and an output layer [37,38]. The hidden layers are vitally important, since they establish the relationship between the outputs and the inputs of the NN, so it is important to define the architecture according to the problem that is being addressed. The hidden layer consists of multiple sublayers connected sequentially, and its depth (i.e., the number of sublayers) depends on the complexity of the problem being addressed. As shown in Figure 1b, the typical structure of a CNN consists of a convolutional layer followed by a pooling layer ("Max Pooling" in Figure 1b), the final layer being a fully connected layer, which is connected to a classification layer in order to output values of either 0 or 1, which define the pertinence class of the input sample.

The convolutional layer is a fundamental part of the neural network, because it extracts the relevant information from the input data [39] by applying the convolution operation between the input matrix and a filter of a given size. The output is known as the feature map. The complexity of the convolutional layer depends on the number of filters and their size. The last component of the convolutional layer is the activation function, which is applied to the feature map to avoid convergence problems. To this end, it applies a nonlinear transformation by means of an activation function, such as the sigmoidal, hyperbolic tangent, rectified linear unit (ReLU), or Maxout functions. The pooling layer often follows the convolutional layer. It is applied to reduce the size of the data coming from the convolutional layer, thus reducing the computational effort and improving the classification accuracy [39]. Finally, the fully connected layer is where the classification process takes place. Neurons in a fully connected layer are connected to all activations in the previous layer. The fully connected layer compiles the data extracted by the preceding layers to generate the final output.

A CNN requires a specific data structure, so it is necessary to accommodate the input data to the requirements of the CNN. Therefore, the input data are fed to the CNN as a matrix, where the columns are the wavelengths of the spectral data, while each row corresponds to a different sample. The output of the CNN is a classification matrix, where every column corresponds to the probability of the sample belonging to each of the defined classes, and every row refers to a different sample.

The next step is to define the structure of the CNN. In this case, two convolutional layers with the subsequent pooling layers were used, as shown in Figure 1b, as a compromise between computational effort and accuracy. The fully connected layer has as many outputs as the number of classes defined in the problem. In this case, the fully connected layer was followed by a classification layer to ensure that only one of the output values was 1, with the others being 0.

2.5. Training of the CNN

Two convolutional layers were selected after training CNNs with different numbers of layers, this solution offering a good balance between computational burden and accuracy [40]. The Adam solver was selected because it offers accurate results, since it efficiently updates the neural network weights [41]. The results of the training process are also highly dependent on the values of the different parameters of the CNN. To solve this issue, an optimization approach was considered by applying the Bayesian optimization algorithm (BOA) [42], because this requires fewer iterations in comparison with traditional optimization algorithms. The BOA initially assumes a prior distribution model of the parameters' function, and uses the data obtained to optimize the training model. The BOA finds the values of the parameters in a bounded domain to improve the result by finding a global optimum (which minimizes the training error) using the information of the previous analyzed point [43]. The BOA was applied to determine the optimal values of the number of epochs (NE), the validation frequency (VF), the validation patience (VP), the verbose frequency (VFreq), the number of filters in the first layer (NF_1st) and the number of filters in the second layer (NF_2nd) during the training process. The BOA function implemented in MATLAB® automatically stops when reaching any of the three following conditions: reaching a fixed number of iterations (30 by default), a fixed time (no time limit by default), or a stopping criterion.

3. Sample Collection and Identification

This paper analyzes textile fibers obtained from several catalogs of different companies. The analyzed samples (textile fibers) included natural, synthetic, and artificial fibers, as well as combinations of natural and synthetic/artificial fibers in different proportions. Natural fibers include silk, wool, linen, and cotton. Synthetic fibers include polyamide and polyester, while viscose is an artificial fiber. While artificial fibers are generated by transforming natural products (e.g., viscose comes from cellulose), synthetic fibers are generated from petrochemical polymers. Hereafter, artificial and synthetic fibers will be referred to as manmade fibers.

A sequence of three studies was designed. For each study, the selected textile fibers were chosen to include maximum variability, so different presentations (fabric or yarn) and a multitude of colors (from dark to light) were included in the analyzed set of samples. The three studies are described below (see Table 1):

- Study #1: To classify 100% pure textile samples among seven different classes (i.e., cotton, linen, wool, silk, polyester, polyamide, and viscose).
- Study #2: To classify mixtures of viscose and polyester in different percentages.
- Study #3: To classify mixtures of cotton and polyester in different percentages.

Samples were provided by different companies (a total of eight) in a span of 4 years (2016 to 2019). A total of 52 different commercial catalogues together with additional spare samples provided by the companies was used to obtain the resulting samples for each of the three studies. The number of catalogues used in each study is presented in Table 1, as well as the number of samples per class and per study. The total number of samples used (considering the three studies) was just above 300. Samples could not be classified visually because their appearance varied widely. The composition of each sample is assured because it is exactly defined in the catalogue provided by the company.

Table 1. Origin and number of samples used in each study.

Studies	Catalogs (n°)	Classification Classes	Samples per Class	Samples per Study
Study #1	25	Cotton; linen; wool; silk; polyester; polyamide; viscose	30	210
Study #2	11	Viscose (100%); viscose/PE (90%/10%); viscose/PE (70–75%/30–25%)	26 26 21	73
Study #3	25	Cotton (>97%); cotton/PE (70–90%/30–10%); cotton/PE (30–65%/70–35%)	30 30 30	90

4. The Analyzed Spectral Data

This study is based on the NIR spectral data of the analyzed textile fibers. To acquire the NIR spectra of the studied samples, an NIR spectrometer equipped with a reflectance fiber-optic module (FOSS XDS^{TM} OptiProbe Analyzer provided by FOSS AnalyticalA/S, Hillerod, Denmark.) controlled by the Vision Software^{TM} was used. Although this was able to acquire the spectra in the 400–2500 nm range, the 1100–2200 nm range was analyzed to reduce the effect of noise and to avoid the visible region. To further reduce the effect of noise, the spectral data were acquired through averaging of 32 scans. Since the resolution of the FOSS XDS^{TM} OptiProbe Analyzer is 0.5 nm, each spectrum has 2201 points per sample. Next, the raw data provided by the NIR spectrometer were processed and transformed by means of the mathematical methods detailed in Section 2.

The NIR spectra of some of the samples analyzed are shown in the following figures. Figure 2a shows the NIR spectra of natural fibers (i.e., cotton, linen, wool, and silk), while manmade fibers are presented in Figure 2b. The similarities and differences between those samples in different ranges of the NIR spectrum can be observed (i.e., the similarity between cotton and linen is high, with both of them also being similar to viscose, while the synthetic fibers are very different from them).

There were three main characteristic bands for cotton: one around 1480 nm, the second at 1942 nm, and the third at 2100 nm. A similar pattern was observed for linen and viscose, as both are cellulose-based fibers, like cotton. In the case of polyamide, a peculiar band was observed at 1713 nm, which corresponds to combination overtones of NH bonds from the amide functional group [44]. This band was also observed in wool and silk (due to amide functional groups from proteins). Finally, the most distinctive band for polyester was found at 1661 nm.

In addition, spectra from binary mixtures of fibers (viscose/PE and cotton/PE) are shown in Figure 2c,d, respectively. In this case, the differences are more difficult to identify at first sight. As noted above, the NIR spectra exhibit wide bands, overlapped bands, and overtones, making it difficult to interpret the spectral information. Therefore, mathematical treatment of the spectra is needed. Nevertheless, we can clearly observe in Figure 2d that, when increasing the amount of PE in the binary blend, the characteristic band of this fiber (at 1661 nm) grows.

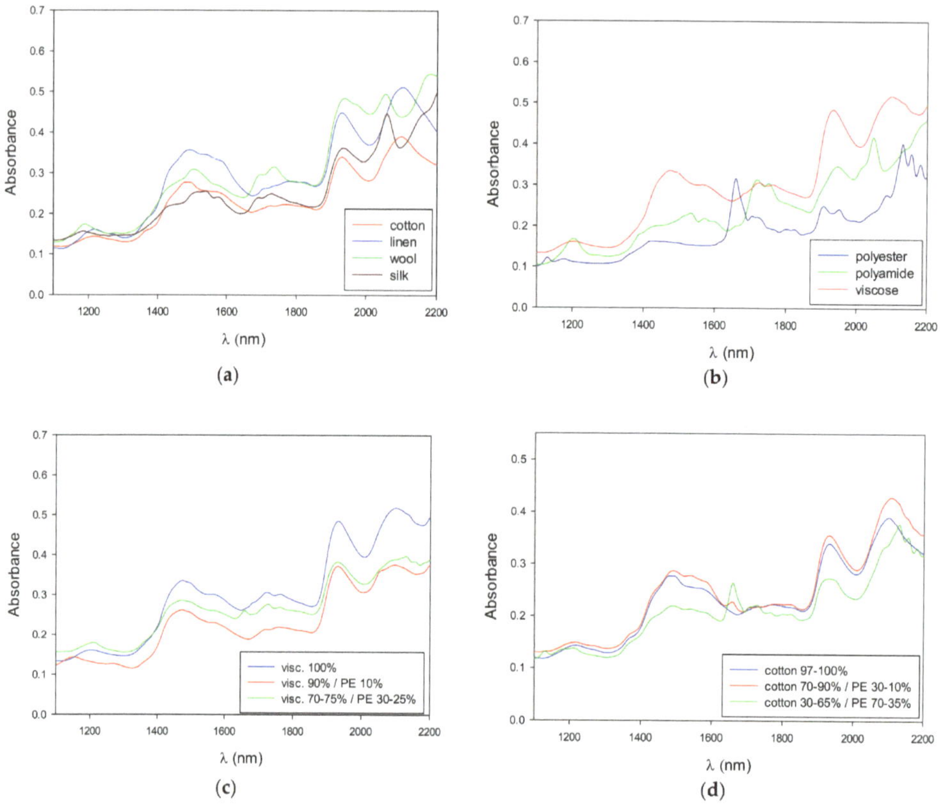

Figure 2. (**a**) Natural fibers. (**b**) Manmade fibers. (**c**) NIR spectra of binary mixtures (viscose/PE). (**d**) NIR spectra of binary mixtures (cotton/PE).

5. Experimental Results

In this section, three studies with different textile samples are presented to prove the accuracy of the proposed approach.

5.1. Study #1: Analyzing Pure Textile Fibers

This first study analyzed 210 pure textile samples, i.e., the samples were made of pure fibers, as summarized in Table 1.

As shown in Table 2, Study #1 considered seven classes of fibers; thus, the goal was to correctly classify all samples of the prediction set, which was composed of $15 \times 7 = 105$ samples. Two approaches were assessed to this end: the first was based on directly applying the CNN to the raw spectral data, or to the first or second derivatives, and mean-centering the data; whereas the second alternative applied the PCA + *CVA* sequence prior to the CNN, as described in Figure 1.

Table 2 summarizes the classification results attained with the two approaches, as well as the CNN and PCA parameters selected in each test. These parameters are the number of epochs (NE), validation frequency (VF), validation patience (VP), verbose frequency (VFreq), the number of filters in the first layer (NF_1st), and the number of filters in the second layer (NF_2nd). Finally, we retained a number of PCs explaining more than 99.99% of the total variance.

The results in Table 3 clearly show the better performance of the PCA + CVA + CNN approach compared to the performance of the CNN alone. This implies that the PCA + CVA feature-selection stage optimally prepares the data for the classification phase, thus simplifying the task of the classifier, greatly reducing the dimensionality of the problem and the computational burden during the training stage, since the dimensionality was reduced from 2201 variables to only 6 in this study (i.e., the number of classes minus one). Therefore, the subsequent studies were based only on PCA + CVA + CNN, due to the superior results obtained by means of this approach. The results summarized in Table 3 also prove that the PCA + CVA + CNN approach enables us to correctly classify the 105 samples of the prediction set.

Table 2. Study #1: Pure textile samples.

Pure Fiber	Type of Fiber	Samples Number		
		Calibration	Prediction	Total
Cotton	Natural	15	15	30
Linen	Natural	15	15	30
Wool	Natural	15	15	30
Silk	Natural	15	15	30
Polyester (PE)	Synthetic	15	15	30
Polyamide (PA)	Synthetic	15	15	30
Viscose	Artificial	15	15	30

Table 3. Study #1: Prediction set classification errors (105 samples).

Conditions	Classification Errors	
	CNN	PCA + CVA + CNN
Mean-centering	4/105 [a]	0/105 [aa]
First derivative + mean-centering	3/105 [b]	0/105 [bb]
Second derivative + mean-centering	4/105 [c]	1/105 [cc]

[a] NE = 200, VF = 8, VP = 29, VFreq = 10, NF_1st = 8, NF_2nd = 16. [b] NE = 500, VF = 18, VP = 20, VFreq = 10, NF_1st = 9, NF_2nd = 14. [c] NE = 385, VF = 18, VP = 20, VFreq = 10, NF_1st = 9, NF_2nd = 12. [aa] NE = 300, VF = 1, VP = 30, VFreq = 1, NF_1st = 8, NF_2nd = 16, 14 PCs. [bb] NE = 600, VF = 9, VP = 10, VFreq = 9, NF_1st = 6, NF_2nd = 18, 45 PCs. [cc] NE = 500, VF = 24, VP = 30, VFreq = 28, NF_1st = 10, NF_2nd = 16, 64 PCs.

5.2. Study #2: Analyzing Mixed Viscose–Polyester Textile Samples

This second study analyzed 73 mixed textile samples, which included pure viscose and viscose/PE mixtures, as summarized in Table 4.

Table 4. Study #2: Pure viscose and viscose/PE mixtures.

Pure Fiber	Composition	Samples Number		
		Calibration	Prediction	Total
Viscose	100% (Pure)	13	13	26
Viscose/PE	90%/10%	13	13	26
Viscose/PE	70–75%/30–25%	10	11	21

As shown in Table 4, Study #2 considered three classes of fibers; thus, the goal was to correctly classify all samples of the prediction set, which was composed of $13 \times 2 + 10 = 36$ samples. Due to the superior results, Study #2 only contemplated the PCA + CVA + CNN approach.

Table 5 summarizes the classification results attained with the PCA + CVA + CNN approach, as well as the CNN and PCA parameters selected in each test.

Table 5. Study #2: Prediction set classification errors (36 samples).

Conditions	Classification Errors
	PCA + CVA + CNN
Mean-centering	1/36 [a]
First derivative + mean-centering	0/36 [b]
Second derivative + mean-centering	0/36 [c]

[a] NE = 605, VF = 1, VP = 13, VFreq = 7, NF_1st = 5, NF_2nd = 1, 8 PCs. [b] NE = 200, VF = 4, VP = 15, VFreq = 4, NF_1st = 6, NF_2nd = 18, 27 PCs. [c] NE = 327, VF = 22, VP = 12, VFreq = 7, NF_1st = 6, NF_2nd = 12, 32 PCs.

As seen in Table 5, when applying the first and second derivatives to the spectral data, the PCA + CVA + CNN method correctly classifies the 36 samples of the prediction set.

5.3. *Study #3: Analyzing Mixed Cotton–Polyester Textile Samples*

This third study analyzed 90 mixed textile samples, which included pure cotton and cotton/PE mixtures, as summarized in Table 6.

Table 6. Study #3: Pure cotton and cotton/PE mixtures.

Pure Fiber	Composition	Samples Number		
		Calibration	Prediction	Total
Cotton	≥97% (Pure)	15	15	30
Cotton/PE	70–90%/30–10%	15	15	30
Cotton/PE	30–65%/70–35%	15	15	30

As shown in Table 6, Study #3 considered three classes of fibers; thus, the goal was to correctly classify all samples of the prediction set, which was composed of 15 × 3 = 45 samples. Due to the superior results, Study #3 only contemplated the PCA + CVA + CNN approach.

Table 7 summarizes the classification results attained with the PCA + CVA + CNN approach, as well as the CNN and PCA parameters selected in each test.

Table 7. Study #3: Prediction set classification errors (36 samples).

Conditions	Classification Errors
	PCA + CVA + CNN
Mean-centering	7/45 [a]
First derivative + mean-centering	7/45 [b]
Second derivative + mean-centering	4/45 [c]

[a] NE = 500, VF = 21, VP = 28, VFreq = 5, NF_1st = 8, NF_2nd = 16, 12 PCs. [b] NE = 300, VF = 7, VP = 6, VFreq = 10, NF_1st = 8, NF_2nd = 16, 27 PCs. [c] NE = 404, VF = 4, VP = 29, VFreq = 23, NF_1st = 8, NF_2nd = 16, 42 PCs.

As seen in Table 7, the PCA + CVA + CNN method correctly classifies between 38 and 41 out of the 45 samples of the prediction set. Some of the misclassified samples are used to make shoes, thus presenting a particular finishing type.

6. Conclusions

The main conclusion of the present work is that the developed method (using a near-infrared (NIR) sensor and mathematical treatment based on convolutional neural networks) allows us to correctly classify pure textile fibers (i.e., cotton, linen, wool, silk, polyester, polyamide, and viscose) and binary blends (i.e., viscose/polyester or cotton/polyester).

The results in Table 3 clearly show the better performance of the PCA + CVA + CNN approach compared to the performance of the CNN alone. This implies that the PCA + CVA feature-selection stage optimally prepares the data for the classification phase, thus simplifying the task of the classifier, greatly reducing the dimensionality of the problem and the computational burden during the training stage, since the dimensionality is reduced from 2201 variables to only 6 or 2 in this study (i.e., the number of classes minus one).

The best classification was obtained with the consecutive application of the PCA + CVA + kNN methods, achieving 100% correct classification rates for pure samples in absorbance mode, 100% correct classification rates for viscose/polyester blends (viscose/PE) in first derivative mode, and 91.1% for cotton/polyester blends (cotton/PE) in second derivative mode.

It should be remembered that in the latter case the three classes defined for the cotton/polyester blends (cotton/PE) were very broad and similar: cotton > 97%; cotton 70–90%; and cotton 30–65%. In contrast, the viscose/PE mixtures were three narrower groups, and were more separated from one another: viscose 100%; viscose 90%; and viscose 70–75%.

The method is powerful, and can be trained to separate pure fibers and the most interesting binary mixtures; its only limitation is the need for a sufficiently large database with samples of known composition. This can help the textile industry to be more circular, by obtaining secondary textile polymers and reducing the need for natural (i.e., crop- or animal-based) and synthetic (i.e., oil-based) virgin materials.

Author Contributions: Conceptualization, J.-R.R., R.C. and R.P.; methodology, J.-R.R. and P.R.-M.; validation, J.-R.R. and P.R.-M.; formal analysis, J.-R.R. and R.P.; investigation, R.C. and R.P.; writing—original draft preparation, J.-R.R. and R.P.; writing—review and editing, J.-R.R. and R.P. All authors have read and agreed to the published version of the manuscript.

Funding: This research was partially funded by the Ministerio de Industria, Comercio, y Turismo de España under grant number AEI-010400-2020-206, and by the Generalitat de Catalunya, under grant numbers 2017 SGR 967 and 2017 SGR 828.

Institutional Review Board Statement: Not applicable.

Informed Consent Statement: Informed consent was obtained from all subjects involved in the study.

Data Availability Statement: Not applicable.

Acknowledgments: The authors are grateful to FITEX and MODACC (textile associations), participant and coordinator respectively of the abovementioned project (AEI), for their technical support and for providing the samples analyzed in this paper.

Conflicts of Interest: The authors declare no conflict of interest.

References

1. Ellen MacArthur Foundation a New Textiles Economy: Redesigning Fashion's Future. Available online: https://emf.thirdlight.com/link/2axvc7eob8zx-za4ule/@/preview/1?o (accessed on 16 May 2022).
2. European Environmental Agency Textiles in Europe's Circular Economy Key Messages. Available online: https://www.eea.europa.eu/publications/textiles-in-europes-circular-economy (accessed on 16 May 2022).
3. Kleinhückelkotten, S.; Neitzke, H.-P. Social Acceptability of More Sustainable Alternatives in Clothing Consumption. *Sustainability* **2019**, *11*, 6194. [CrossRef]
4. Hole, G.; Hole, A.S. Recycling as the way to greener production: A mini review. *J. Clean. Prod.* **2019**, *212*, 910–915. [CrossRef]
5. Yousef, S.; Tatariants, M.; Tichonovas, M.; Sarwar, Z.; Jonuškienė, I.; Kliucininkas, L. A new strategy for using textile waste as a sustainable source of recovered cotton. *Resour. Conserv. Recycl.* **2019**, *145*, 359–369. [CrossRef]
6. Terinte, N.; Manda, B.; Taylor, J.; Schuster, K.; Patel, M. Environmental assessment of coloured fabrics and opportunities for value creation: Spin-dyeing versus conventional dyeing of modal fabrics. *J. Clean. Prod.* **2014**, *72*, 127–138. [CrossRef]
7. Roos, S.; Jönsson, C.; Posner, S.; Arvidsson, R.; Svanström, M. An inventory framework for inclusion of textile chemicals in life cycle assessment. *Int. J. Life Cycle Assess.* **2019**, *24*, 838–847. [CrossRef]
8. Piribauer, B.; Bartl, A. Textile recycling processes, state of the art and current developments: A mini review. *Waste Manag. Res. J. A Sustain. Circ. Econ.* **2019**, *37*, 112–119. [CrossRef]
9. Esteve-Turrillas, F.; de la Guardia, M. Environmental impact of Recover cotton in textile industry. *Resour. Conserv. Recycl.* **2017**, *116*, 107–115. [CrossRef]
10. Sandin, G.; Peters, G.M. Environmental impact of textile reuse and recycling—A review. *J. Clean. Prod.* **2018**, *184*, 353–365. [CrossRef]
11. De Oliveira, C.R.S.; Júnior, A.H.D.S.; Mulinari, J.; Immich, A.P.S. Textile Re-Engineering: Eco-responsible solutions for a more sustainable industry. *Sustain. Prod. Consum.* **2021**, *28*, 1232–1248. [CrossRef]

12. Dahlbo, H.; Aalto, K.; Eskelinen, H.; Salmenperä, H. Increasing textile circulation—Consequences and requirements. *Sustain. Prod. Consum.* **2017**, *9*, 44–57. [CrossRef]
13. Huang, Y.-F.; Azevedo, S.G.; Lin, T.-J.; Cheng, C.-S.; Lin, C.-T. Exploring the decisive barriers to achieve circular economy: Strategies for the textile innovation in Taiwan. *Sustain. Prod. Consum.* **2021**, *27*, 1406–1423. [CrossRef]
14. Hole, G.; Hole, A.S. Improving recycling of textiles based on lessons from policies for other recyclable materials: A minireview. *Sustain. Prod. Consum.* **2020**, *23*, 42–51. [CrossRef]
15. *Document 32018L0851*; Directive (EU) 2018/851 of the European Parliament and of the Council of 30 May 2018 Amending Directive 2008/98/EC on Waste (Text with EEA Relevance). Official Journal of the European Union, 2018. Available online: https://eur-lex.europa.eu/eli/dir/2018/851/oj (accessed on 16 May 2022).
16. Nørup, N.; Pihl, K.; Damgaard, A.; Scheutz, C. Evaluation of a European textile sorting centre: Material flow analysis and life cycle inventory. *Resour. Conserv. Recycl.* **2019**, *143*, 310–319. [CrossRef]
17. Nunes, M.A.; Páscoa, R.N.; Alves, R.C.; Costa, A.S.; Bessada, S.; Oliveira, M.B.P. Fourier transform near infrared spectroscopy as a tool to discriminate olive wastes: The case of monocultivar pomaces. *Waste Manag.* **2020**, *103*, 378–387. [CrossRef]
18. Salzmann, M.; Blößl, Y.; Todorovic, A.; Schledjewski, R. Usage of Near-Infrared Spectroscopy for Inline Monitoring the Degree of Curing in RTM Processes. *Polymers* **2021**, *13*, 3145. [CrossRef]
19. Chen, X.; Kroell, N.; Wickel, J.; Feil, A. Determining the composition of post-consumer flexible multilayer plastic packaging with near-infrared spectroscopy. *Waste Manag.* **2021**, *123*, 33–41. [CrossRef]
20. Huang, F.; Song, H.; Guo, L.; Guang, P.; Yang, X.; Li, L.; Zhao, H.; Yang, M. Detection of adulteration in Chinese honey using NIR and ATR-FTIR spectral data fusion. *Spectrochim. Acta Part A Mol. Biomol. Spectrosc.* **2020**, *235*, 118297. [CrossRef]
21. Cruz-Tirado, J.; Medeiros, M.L.D.S.; Barbin, D.F. On-line monitoring of egg freshness using a portable NIR spectrometer in tandem with machine learning. *J. Food Eng.* **2021**, *306*, 110643. [CrossRef]
22. Corrêdo, L.P.; Wei, M.C.; Ferraz, M.N.; Molin, J.P. Near-infrared spectroscopy as a tool for monitoring the spatial variability of sugarcane quality in the fields. *Biosyst. Eng.* **2021**, *206*, 150–161. [CrossRef]
23. Wang, Z.; Peng, B.; Huang, Y.; Sun, G. Classification for plastic bottles recycling based on image recognition. *Waste Manag.* **2019**, *88*, 170–181. [CrossRef]
24. Riba, J.-R.; Cantero, R.; García-Masabet, V.; Cailloux, J.; Canals, T.; Maspoch, M.L. Multivariate identification of extruded PLA samples from the infrared spectrum. *J. Mater. Sci.* **2020**, *55*, 1269–1279. [CrossRef]
25. De Maesschalck, R.; Estienne, F.; Verdu-Andres, J.; Candolfi, A.; Centner, V.; Despagne, F.; Jouan-Rimbaud, D.; Walczak, B.; Massart, D.L.; De Jong, S.; et al. The development of calibration models for spectroscopic data using principal component regression. *Internet J. Chem.* **1999**, *2*.
26. Riba, J.-R.; Cantero, R.; Canals, T.; Puig, R. Circular economy of post-consumer textile waste: Classification through infrared spectroscopy. *J. Clean. Prod.* **2020**, *272*, 123011. [CrossRef]
27. Pan, J.; Nguyen, K.L. Development of the Photoacoustic Rapid-Scan FT-IR-Based Method for Measurement of Ink Concentration on Printed Paper. *Anal. Chem.* **2007**, *79*, 2259–2265. [CrossRef]
28. Riba, J.-R.; Canals, T.; Cantero, R. Recovered Paperboard Samples Identification by Means of Mid-Infrared Sensors. *IEEE Sens. J.* **2013**, *13*, 2763–2770. [CrossRef]
29. Nørgaard, L.; Bro, R.; Westad, F.; Engelsen, S.B. A modification of canonical variates analysis to handle highly collinear multivariate data. *J. Chemom.* **2006**, *20*, 425–435. [CrossRef]
30. Lai, W.-W.; Zeng, X.-X.; He, J.; Deng, Y.-L. Aesthetic defect characterization of a polymeric polarizer via structured light illumination. *Polym. Test.* **2016**, *53*, 51–57. [CrossRef]
31. Younes, K.; Moghrabi, A.; Moghnie, S.; Mouhtady, O.; Murshid, N.; Grasset, L. Assessment of the Efficiency of Chemical and Thermochemical Depolymerization Methods for Lignin Valorization: Principal Component Analysis (PCA) Approach. *Polymers* **2022**, *14*, 194. [CrossRef]
32. Riba Ruiz, J.-R.; Canals, T.; Cantero, R. Supervision of Ethylene Propylene Diene M-Class (EPDM) Rubber Vulcanization and Recovery Processes Using Attenuated Total Reflection Fourier Transform Infrared (ATR FT-IR) Spectroscopy and Multivariate Analysis. *Appl. Spectrosc.* **2017**, *71*, 141–151. [CrossRef]
33. Bhattacharyya, N.; Bandyopadhyay, R.; Bhuyan, M.; Tudu, B.; Ghosh, D.; Jana, A. Electronic Nose for Black Tea Classification and Correlation of Measurements With "Tea Taster" Marks. *IEEE Trans. Instrum. Meas.* **2008**, *57*, 1313–1321. [CrossRef]
34. Amor, N.; Noman, M.T.; Petru, M. Classification of Textile Polymer Composites: Recent Trends and Challenges. *Polymers* **2021**, *13*, 2592. [CrossRef]
35. Cun, L.; Henderson, J.; Le Cun, Y.; Denker, J.S.; Henderson, D.; Howard, R.E.; Hubbard, W.; Jackel, L.D. Handwritten Digit Recognition with a Back-Propagation Network. In *Advances in Neural Information Processing Systems*; Touretzk, D., Ed.; Morgan-Kaufmann: Cambridge, MA, USA, 1989; Volume 2, pp. 396–404.
36. Zhang, Q.; Yang, Q.; Zhang, X.; Bao, Q.; Su, J.; Liu, X. Waste image classification based on transfer learning and convolutional neural network. *Waste Manag.* **2021**, *135*, 150–157. [CrossRef]
37. Cao, X.-C.; Chen, B.-Q.; Yao, B.; He, W.-P. Combining translation-invariant wavelet frames and convolutional neural network for intelligent tool wear state identification. *Comput. Ind.* **2019**, *106*, 71–84. [CrossRef]
38. Ciancetta, F.; Bucci, G.; Fiorucci, E.; Mari, S.; Fioravanti, A. A New Convolutional Neural Network-Based System for NILM Applications. *IEEE Trans. Instrum. Meas.* **2020**, *70*, 1–12. [CrossRef]

39. Rawat, W.; Wang, Z. Deep convolutional neural networks for image classification: A comprehensive review. *Neural Comput.* **2017**, *29*, 2352–2449. [CrossRef]
40. Rojas-Duenas, G.; Riba, J.-R.; Moreno-Eguilaz, M. Black-Box Modeling of DC–DC Converters Based on Wavelet Convolutional Neural Networks. *IEEE Trans. Instrum. Meas.* **2021**, *70*, 1–9. [CrossRef]
41. Ruder, S. An Overview of Gradient Descent Optimization Algorithms. 2016. Available online: https://arxiv.org/abs/1609.04747 (accessed on 16 May 2022).
42. He, F.; Zhou, J.; Feng, Z.-K.; Liu, G.; Yang, Y. A hybrid short-term load forecasting model based on variational mode decomposition and long short-term memory networks considering relevant factors with Bayesian optimization algorithm. *Appl. Energy* **2019**, *237*, 103–116. [CrossRef]
43. Shahriari, B.; Swersky, K.; Wang, Z.; Adams, R.P.; De Freitas, N. Taking the human out of the loop: A review of Bayesian optimization. *Proc. IEEE* **2016**, *104*, 148–175. [CrossRef]
44. Sun, X.; Zhou, M.; Sun, Y. Classification of textile fabrics by use of spectroscopy-based pattern recognition methods. *Spectrosc. Lett.* **2016**, *49*, 96–102. [CrossRef]

polymers

MDPI

Article

Comparative Evaluation of Locally Administered 2% Gel Fabricated from Lemongrass Polymer and 10% Doxycycline Hyclate Gel as an Adjunct to Scaling and Root Planing in the Treatment of Chronic Periodontitis—A Randomized Controlled Trial

Pooja Mittal [1], Shankar T. Gokhale [2], Shiva Manjunath [3], Saad M. Al-Qahtani [2], Mohammad Al. Magbol [2], Raghavendra Reddy Nagate [2], Shreyas Tikare [2], Saurabh Chaturvedi [4,*], Ashish Agarwal [5] and Vatsala Venkataram [6]

1 Department of Periodontology, Dental College Azamgarh, Azamgarh 276128, Uttar Pradesh, India; drmittalpooja@gmail.com

2 Department of Periodontics and Community Dental Sciences, College of Dentistry, King Khalid University, Abha 61421, Saudi Arabia; sgokhale@kku.edu.sa (S.T.G.); saad71@outlook.in (S.M.A.-Q.); moama1957@outlook.com (M.A.M.); rnagati@kku.edu.sa (R.R.N.); tikane@kku.edu.sa (S.T.)

3 Department of Periodontics, Institute of Dental Sciences, Bareilly 243006, Uttar Pradesh, India; drmanju75@rediffmail.com

4 Department of Prosthetic Dentistry, College of Dentistry, King Khalid University, Abha 61421, Saudi Arabia

5 Community Health Center, Puranpur, Pilibhit 262121, Uttar Pradesh, India; drashish.aag@gmail.com

6 Department of Pedodontics & Preventive Dentistry, KVG Dental College & Hospital, Sullia 574327, Karnataka, India; dr.vatsala@gmail.com

* Correspondence: survedi@kku.edu.sa; Tel.: +966-580-697248

Abstract: Background: Extracts of medicinal plant like lemongrass offer a new choice for optional antimicrobial therapy against various oral microorganisms. The objective of this study was to assess, verify, and compare the antimicrobial effectiveness of locally administered 2% lemongrass gel and 10% doxycycline hyclate gel as an adjunct to scaling and root planing (SRP) in treating chronic periodontitis. Method: This is a double-blind parallel arm randomized controlled study. Forty subjects were randomly divided into Group A and B for 2% lemongrass gel and 10% doxycycline hyclate gel, respectively. The clinical assessments of Gingival Index (GI), Plaque Index (PI), Probing Pocket Depth (PPD), and Clinical Attachment Level (CAL) together with microbial colony counts for *Porphyromonas gingivalis*, *Actinomyces naeslundii*, and *Prevotella intermedia* were done at baseline, 1st month, and 3rd month follow-ups. Results: The results showed there was a significant reduction in the mean scores of GI, PPD, and CAL clinical indices from baseline to the 1st and 3rd month follow-ups in both the 2% lemongrass gel and 10% doxycycline gel groups ($p < 0.05$). Similarly, there was significant reduction in mean CFU scores for all periodontal pathogens from baseline to 1st and 3rd month follow-ups in both the 2% lemongrass gel and 10% doxycycline gel groups ($p < 0.05$). Conclusions: It could be concluded that the local delivery of 2% lemongrass gel as an adjunct to scaling and root planing is effective and comparable to 10% doxycycline gel in the treatment of chronic periodontitis.

Keywords: cymbopogon citratus; local drug delivery; non-surgical therapy; periodontal pathogens

1. Introduction

The applications of polymers in dentistry are increasing day by day. The irreplaceable merits of different classes of natural polymers over synthetic polymers as drug carriers still urge researchers to depend solely on biomaterials. These advantages include hydrophilicity, biocompatibility, non-immunogenicity, non-toxicity, non-toxic degradation products, antimicrobial and antioxidant capacities, and high stability for tissue engineering [1–3]. Also, from an environmental perspective, the worldwide demand for eco-friendly products has witnessed sustainable growth for accommodating green technology advancements. Hence, industrial channels are searching for more reliable products to gain patients' satisfaction and ease their commercialization [4–8].

One such field is pain relief in dentistry. The use of polymers in dentistry has evolved from simple reinforcement for dental materials to application in healing and pain relief. Periodontitis is defined as 'an inflammatory disease of the supporting tissues of the teeth caused by specific microorganisms or groups of specific microorganisms, resulting in progressive destruction of the periodontal ligament and alveolar bone with increased probing depth formation, recession, or both' [1]. The most prevalent periodontal disease is chronic periodontitis, which is caused by the deposition of bacterial plaque over time. The therapy for gingivitis and periodontitis has evolved quite differently in these two areas, depending on the advancements made in the area of drug and polymer development and changes in patient populations and needs. The advancements that are most pertinent to polymers dealt with (1) their use as carriers for the controlled delivery of bioactive agents, particularly antimicrobials (2) their use in conjunction with tissue regeneration and healing and (3) factors pertinent to the effectiveness of implemented therapies [2,9–14].

Cymbopogon citratus, Stapf, or lemongrass, is a naturally available flavonoid and contains cellulose as a natural polymer. This lemongrass polymer is working towards the quest for the development of a biodegradable natural polymer that has anti-bacterial, anti-filarial, anti-fungal, anti-inflammatory, and antioxidant properties [15]. Lemongrass has active phenol and flavonoid substances which, at concentrations below 2%, have shown to have bacteriostatic properties against several microorganisms [16]. In an in vitro study, lemongrass essential oil was effective against periodontal pathogens like *Actinomyces naeslundii* and *Porphyromonas gingivalis* and also a majority of clinical-isolate groups, including tetracycline hydrochloride-resistant strains [17]. Mouth rinse with active Cymbopogon citratus essential oil as an adjunct to SRP has shown to be effective in reducing the severity of gingivitis [18,19]. The use of lemongrass essential oil as a local drug delivery (LDD) in chronic periodontitis is limited.

Non-surgical mechanical periodontal therapies such as *scaling* and *root planing* (SRP), and in some cases surgical periodontal therapy with access flaps, have been archived widely in the literature to hamper the progression of tissue destruction in periodontal disease [2]. Mechanical therapy, however, may not always reduce or eliminate the anaerobic infection at the base of the pocket, within the gingival tissues, and in areas inaccessible to periodontal instruments [3]. The reduction or elimination of the pathogenic microorganisms in the subgingival microenvironment is indicative of a successful periodontal therapy [1,5].

The adjunctive use of antimicrobial agents in addition to non-surgical therapy has shown to provide additional benefits [6]. The use of systemic antimicrobial agents in the treatment of periodontal disease was widespread in the past. The main drawback of systemic antimicrobials is achieving and maintaining a therapeutic concentration at the infected site. Antimicrobial agents locally applied into the periodontal pockets may further suppress periodontal pathogens. Various local drug delivery (LDD) systems have been introduced to overcome the disadvantages of the systemic route of antimicrobial administration. LDD systems in periodontal therapy can provide up to 100-fold higher drug doses at the target site compared to systemic administration [7]. To date, doxycycline, and metronidazole are the most widely used antibiotics for LDD in the treatment of periodontal disease. The concentration of tetracycline in gingival crevicular fluid (GCF) is 5–10-fold higher than serum levels due to their anti-collagenase property [8,9]. Such antimicrobial therapy as an adjunct requires reduced dosage and fewer applications and also has high patient compliance.

Increasing concern over the unwanted side effects and emergence of highly resistant microbes with increased pathogenicity at the treated sites has altered the general perception of the capabilities of these antimicrobial agents [10,11]. In light of this, there is a need to look for alternate options that are effective, relatively safe, and economical [12]. Research in phytosciences, an emerging multidisciplinary science, has revealed various medicinal plants possessing antimicrobial activity with fewer side effects, reduced toxicity, and cost-effectiveness. Extracts of these medicinal plants offer a new choice for optional antimicrobial therapy against various oral microorganisms [13,14].

Keeping the above factors in mind, the present study was planned with the aim to assess and compare the effectiveness of locally administered 2% gel made from lemongrass polymer and 10% doxycycline hyclate gel as an adjunct to SRP in treating chronic periodontitis and also to verify and compare the antimicrobial effect of 2% gel made from lemongrass polymer and 10% doxycycline hyclate gel in chronic periodontitis.

2. Materials and Methods

2.1. Study Design

A double-blind parallel arm randomized controlled study was planned to evaluate the effectiveness of 2% lemongrass gel as an adjunct to SRP therapy in chronic periodontitis with 10% doxycycline hyclate gel as active control. This study also aimed to evaluate in vivo antimicrobial effect of both LDD agents. The study protocol was developed, and all subjects gave their informed consent for inclusion before they participated in the study. The study was conducted in accordance with the Declaration of Helsinki, and the protocol was approved by the Ethics Committee of Institute of dental science, Bareilly, India. "(IDS/ETHCC/14/10)." The study was registered with the Clinical Trial Registry of India (CTRI REF/2021/03/042330 AU).

Study subjects and Sample size

Sample size (n) was calculated by using formula

$$n = \frac{2(Standard\ deviation)^2}{(Effect\ size)^2}(Z_{\alpha/2} + Z_{1-\beta})^2. \quad (1)$$

where $Z_{\alpha/2} = 1.96$, $Z_{1-\beta} = 0.842$ are respectively the 95% confidence value obtained from the standard normal distribution with power of the study at 80%. At least seventeen subjects were needed to detect a significant difference in PPD after intervention with an effect size of 0.50 and a standard deviation of 0.45 from the pilot study using ten subjects. To compensate for the dropouts, 10% of n is added to get the final sample size. Hence, the final minimum sample size in each group was $n = 20$.

A total of 74 subjects were screened for the study, of which 34 subjects did not satisfy the inclusion criteria. The inclusion criteria for the patient selection were: (a) patients in the age group of 18–60 years, (b) untreated chronic periodontitis with a minimum of four periodontal pockets per quadrant, and (c) periodontal pockets with probing depth 4 to 6 mm. Patients with systemic diseases, who had received antibiotics during the previous three months, or who had received antiseptic/antiplaque agents in the last three months, pregnant women, and patients who had undergone periodontal treatment in the last six months were excluded from the study. The details of study procedure were explained to all eligible subjects ($n = 40$) and consent was obtained.

2.2. Formulation of 2% Lemongrass Gel and 10% Doxycycline Hyclate Gel

For this study, a stable bio-absorbable controlled-release formulation of 2% lemongrass gel and 10% doxycycline hyclate for the treatment of periodontal pockets was prepared by Department of Pharmacy, Mahatma Jyotiba Phule Rohilkhand University, Bareilly, India [20].

2.2.1. Preparation of 2% Lemongrass Gel

Appropriate quantity of Carbopol 934 was soaked in water for a period of 2 h. Carbopol was then neutralized with Triethanolamine (TEA) by stirring. Then, 2% lemongrass essential oil was dissolved in appropriate and pre-weighted amounts of propylene glycol and ethanol. The solvent blend was transferred to the Carbopol container and agitated for an additional 20 min. The dispersion was then allowed to hydrate and swell for 60 mins, then the pH was adjusted with 98% TEA until the desired pH value approximately reached (6.8–7). During pH adjustment, the mixture was stirred gently with a spatula until a homogeneous gel was formed.

2.2.2. Preparation of 10% Doxycycline Hyclate Gel

A total of 5% sesame oil was added to 95% of the melted Glyceryl monooleate (GMO) at 60–70 °C with continuous stirring. After the above solution was cooled to room temperature, 10% of doxycycline hyclate was added to it until a homogenous gel was obtained.

2.3. *Study Procedure*

Forty subjects were randomly allocated into two equal intervention groups using a lottery method with Group A subjects (n = 20) for 2% lemongrass gel and Group B (n = 20) for 10% doxycycline hyclate gel as interventions, respectively. Only one investigator was aware of the intervention groups, and the subjects were coded for identification. The same investigator carried out the placement of local drug delivery for each study subject. The clinical periodontal status of each subject was recorded at baseline, i.e., before SRP therapy and application of local drug delivery and on 1st and 3rd month follow-up assessments. The clinical steps involved are shown in Figure 1.

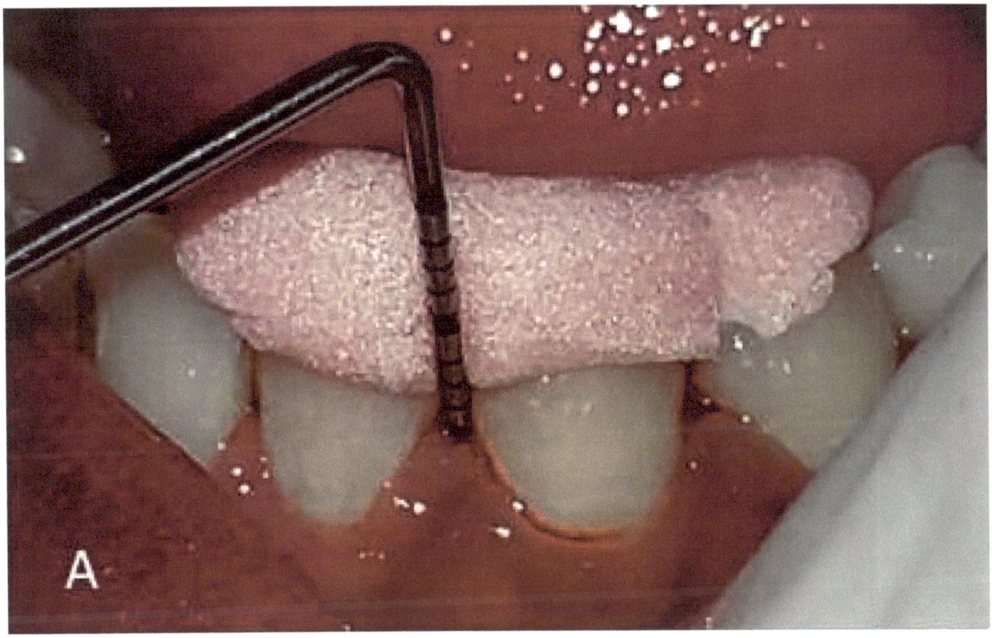

Figure 1. *Cont.*

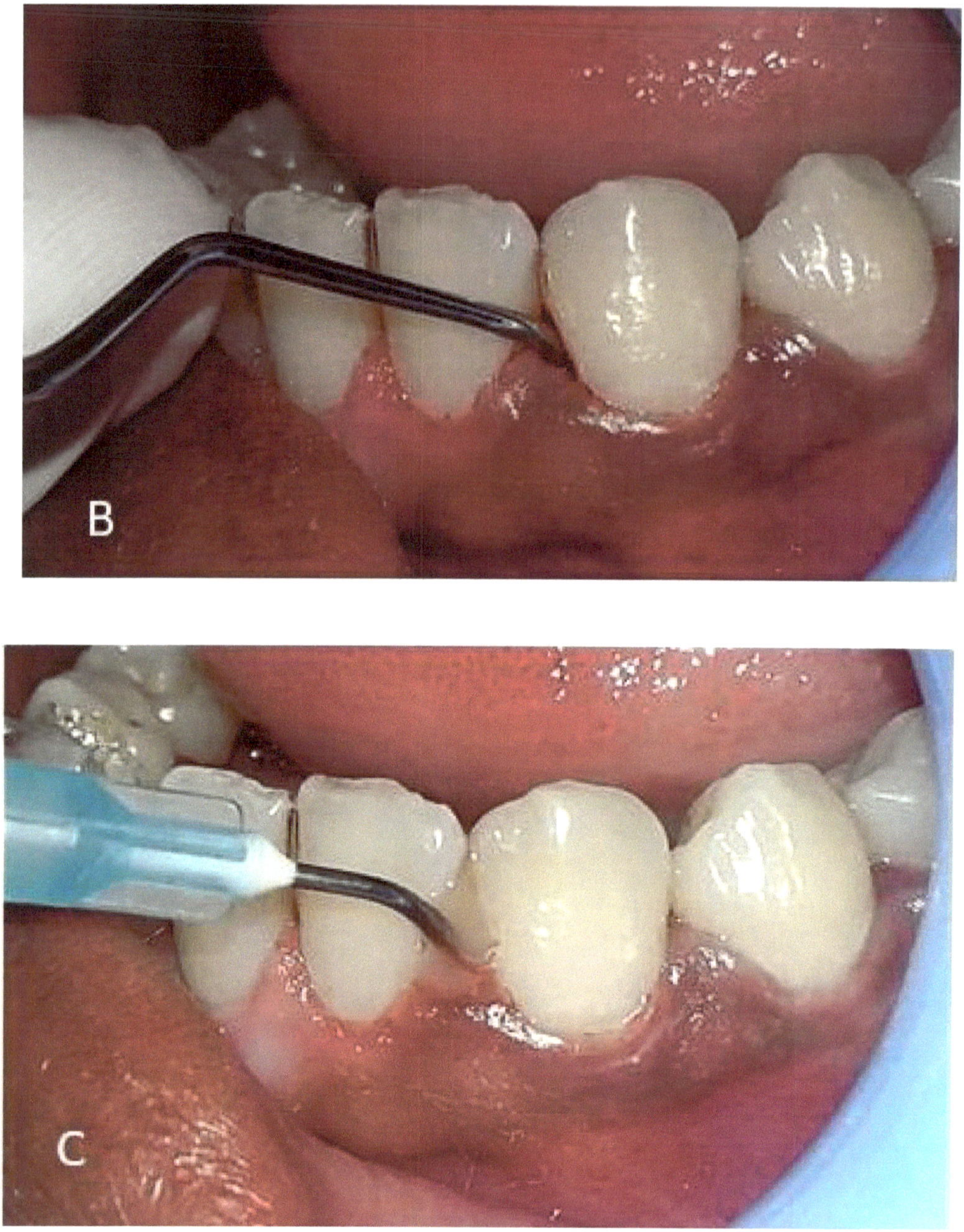

Figure 1. *Cont.*

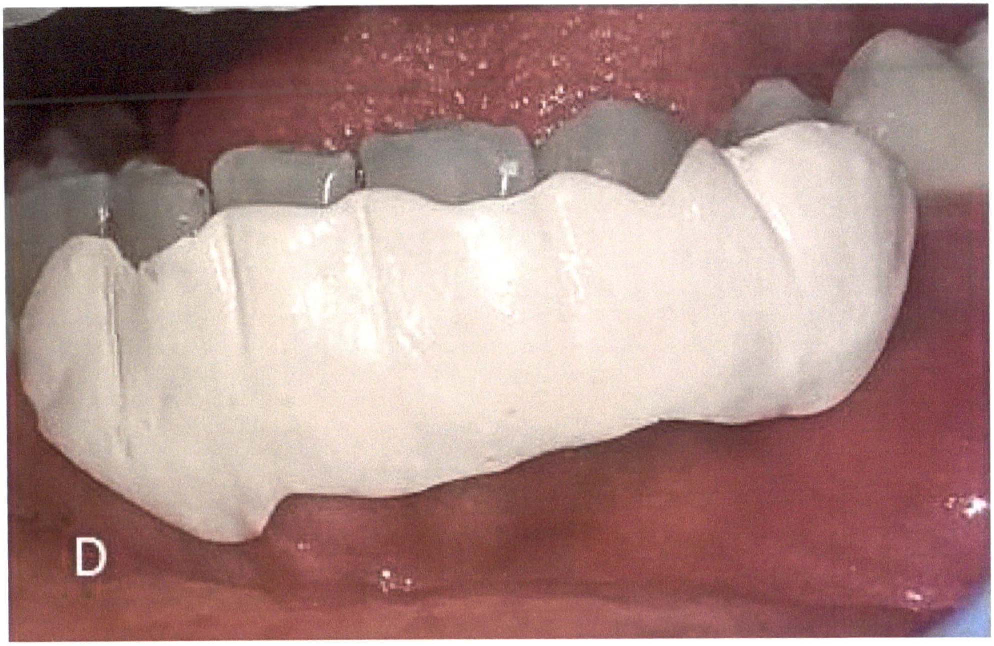

Figure 1. Clinical steps involved in the study. (**A**) Recording clinical parameters at baseline. (**B**) Subgingival plaque sample collection. (**C**) LDD at the site of Periodontal Pocket. (**D**) Placement of periodontal dressing.

In this study both clinical and microbiological periodontal parameters were used for outcomes assessment.

2.4. Clinical Assessments

Clinical assessments for periodontitis were done as per American Academy of Periodontology (AAP) 1999 classification [21]. The indices used were: Gingival index (GI), Plaque Index (PI) Probing Pocket Depth (PPD) and Clinical Attachment Level (CAL) [1,22,23]. All clinical assessments were done with the University of North Carolina-15 probe (Hu Friedy®). A single blinded investigator was trained and calibrated for recording clinical indices with ten independent subjects on an interval of 24 h. The intra-class correlation coefficient was found to be 0.82 which indicate good reliability.

2.5. Microbiological Analysis

Before recording clinical parameters, subgingival plaque samples were collected from the selected sites at baseline, 1st month and 3rd month visits to evaluate the changes in colony forming units (CFU) of primary periodontal pathogens, i.e., *Porphyromonas gingivalis*, *Actinomyces naeslundii* and *Prevotella intermedia* [24]. The teeth were isolated using cotton rolls and a plaque sample was obtained using sterile area specific Gracey curettes (Hu Friedy®) in a previously fumigated minor operation theatre. The samples were then transferred in a vial containing 10 mL Robertson Cooked Meat Broth Medium and were incubated at 37 °C for 24–96 h. The total number of CFU's was determined based on serial dilution from 10–1 to 10–3 on selective media. Finally, each bacteria's count was determined based on typical colony and bacterial morphology in 10^2 CFU/milliliter on Muller Hinton Agar.

2.6. Study Intervention

Immediately after baseline assessment, all subjects received SRP therapy followed by application of local drug delivery. The investigator assigned for application of LDD gels applied 2% lemongrass gel and 10% doxycycline hyclate gel into periodontal pockets as per patient codes. The LDD gel was inserted into the bases of the pockets using a special syringe with a blunt cannula. The end of the blunt cannula was moved coronally to fill the pocket, and the excess gel was removed using a curette or wet cotton pellet. The site was covered with a periodontal dressing (Coe-Pak®) to prevent the medication from being flushed out of the pocket. All subjects were advised to use 0.2% of chlorhexidine rinses, and standard oral hygiene instructions were given. After one week, the subjects were reviewed for any discomfort such as transient discomfort, erythema, transient resistance, allergy following treatment, and visual examination to record any soft tissue changes after removal of periodontal dressing. The subjects were recalled for follow-up assessments on 1st and 3rd months, and oral hygiene maintenance was reinforced. The study design flow chart is given in Figure 2. All subjects who participated completed the study, and none of the subjects had any complications during recall visits.

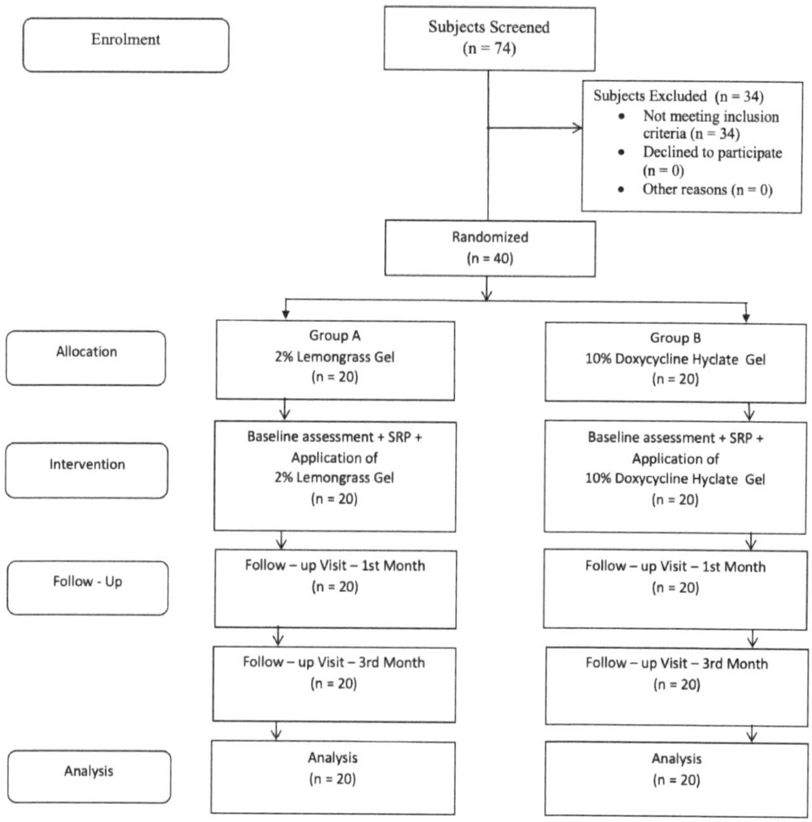

Figure 2. Flow chart of the study.

2.7. Statistical Analysis

The collected data was first entered into MS-Excel spreadsheet and further subjected to analysis using SPSS for Windows, Version 16.0. Chicago, SPSS Inc. (Chicago, IL, USA) Descriptive results were presented as frequency, mean, and standard deviation. Inter-group comparisons were made by independent Student's *t*-test and intra-group comparisons to test changes during follow-up visits were obtained by using repeated measures of ANOVA test and Least Significant Difference post hoc test. The level of significance was set at 5%.

3. Results

At baseline, there was no statistically significant difference between the groups in their mean clinical and microbiological scores ($p > 0.05$) (Table 1). The repeated measures ANOVA test showed that there was a significant reduction in all clinical mean scores from baseline to the 1st and 3rd month follow-ups in both 2% lemongrass gel and 10% doxycycline gel groups ($p < 0.05$) (Table 2). Similarly, there was significant reduction in all microbiological mean CFU scores from baseline to the 1st and 3rd month follow-ups in both 2% lemongrass gel and 10% doxycycline gel groups ($p < 0.05$) (Table 3). A pairwise Least Significant Difference (LSD) post hoc test showed that there was a significant reduction of GI, PPD, CAL, and all three microbiological mean CFU count scores in both the 2% lemongrass gel and 10% doxycycline gel groups from baseline to the 1st and 3rd month follow-up scores ($p < 0.05$). There was no statistically significant difference in the mean PI scores from baseline to the 1st and 3rd month follow-up scores in 2% lemongrass gel group ($p > 0.05$) (Tables 2 and 3).

Table 1. Comparison of mean scores of both clinical and microbiological parameters at baseline.

	Groups	N	Mean	SD	p-Value
GI	2% Lemongrass Gel	20	1.78	0.26	0.868
	10% Doxycycline Gel	20	1.76	0.24	
PI	2% Lemongrass Gel	20	1.82	0.20	0.814
	10% Doxycycline Gel	20	1.84	0.19	
PPD	2% Lemongrass Gel	20	5.30	0.47	0.744
	10% Doxycycline Gel	20	5.35	0.49	
CAL	2% Lemongrass Gel	20	5.50	0.51	0.759
	10% Doxycycline Gel	20	5.55	0.51	
Porphyromonas Gingivalis (10^3 CFU)	2% Lemongrass Gel	20	1.06	0.54	0.757
	10% Doxycycline Gel	20	1.00	0.60	
Actinomyces Naeslundi (10^2 CFU)	2% Lemongrass Gel	20	0.69	0.43	0.683
	10% Doxycycline Gel	20	0.75	0.51	
Prevotella Intermedia (10^3 CFU)	2% Lemongrass Gel	20	1.12	0.48	0.274
	10% Doxycycline Gel	20	0.94	0.52	

Statistically significant at 5% level of significance.

Table 2. Comparison of mean clinical parameters scores within groups at baseline, 1 month, and 3 month follow-ups.

	Groups	Assessment Period	Mean	SD	p-Value
GI	2% Lemongrass Gel	Baseline	1.78	0.26	<0.0001 * Φ
		1 Month	1.34	0.27	
		3 Months	1.22	0.28	
	10% Doxycycline Gel	Baseline	1.76	0.24	<0.0001 * Φ
		1 Month	1.34	0.23	
		3 Months	1.16	0.28	
PI	2% Lemongrass Gel	Baseline	1.82	0.20	0.021 *
		1 Month	1.78	0.25	
		3 Months	1.73	0.23	
	10% Doxycycline Gel	Baseline	1.84	0.19	0.015 * Φ
		1 Month	1.72	0.26	
		3 Months	1.70	0.28	
PPD	2% Lemongrass Gel	Baseline	5.30	0.47	<0.0001 * Φ
		1 Month	3.30	0.47	
		3 Months	3.25	0.44	
	10% Doxycycline Gel	Baseline	5.35	0.49	<0.0001 * Φ
		1 Month	3.40	0.50	
		3 Months	3.15	0.37	
CAL	2% Lemongrass Gel	Baseline	5.50	0.51	<0.0001 * Φ
		1 Month	3.50	0.61	
		3 Months	3.45	0.60	
	10% Doxycycline Gel	Baseline	5.55	0.51	<0.0001 * Φ
		1 Month	3.60	0.50	
		3 Months	3.35	0.49	

* Statistically significant at 5% level of significance by repeated measures ANOVA. Φ Statistically significant at 5% level of significance by pairwise Least Significant Difference post hoc test.

Table 3. Comparison of mean microbiological parameter scores within groups at baseline, 1 month, and 3 month follow-ups.

	Groups	Assessment Period	Mean	SD	*p*-Value
Porphyromonas Gingivalis (10^3 CFU)	2% Lemongrass Gel	Baseline	1.06	0.54	
		1 Month	0.31	0.34	<0.0001 * Φ
		3 Months	0.28	0.31	
	10% Doxycycline Gel	Baseline	1.00	0.60	
		1 Month	0.37	0.33	<0.0001 * Φ
		3 Months	0.38	0.35	
Actinomyces Naeslundi (10^2 CFU)	2% Lemongrass Gel	Baseline	0.69	0.43	
		1 Month	0.33	0.34	<0.0001 * Φ
		3 Months	0.31	0.30	
	10% Doxycycline Gel	Baseline	0.75	0.51	
		1 Month	0.33	0.31	<0.0001 * Φ
		3 Months	0.32	0.31	
Prevotella Intermedia (10^3 CFU)	2% Lemongrass Gel	Baseline	1.12	0.48	
		1 Month	0.70	0.36	<0.0001 * Φ
		3 Months	0.67	0.35	
	10% Doxycycline Gel	Baseline	0.94	0.52	
		1 Month	0.51	0.41	<0.0001 * Φ
		3 Months	0.52	0.41	

* Statistically significant at 5% level of significance by repeated measures ANOVA. Φ Statistically significant at 5% level of significance by pairwise Least Significant Difference post hoc test.

The mean difference was highest in PPD and CAL assessments from baseline to the follow-up visits in both groups. The mean differences of microbial CFUs from baseline to the 1st and 3rd month follow-ups in both groups were comparable (Tables 2 and 3).

4. Discussion

Various polymers have been explained by researchers for drug delivery. Different polymers exhibit different mucoadhesive properties depending on their physical and chemical strength. For example, a more flexible polymer exhibits a higher degree of mucoadhesive property [25]. Mucoadhesive polymers possessing hydrophilic functional groups such as COOH, OH, NH_2, and SO_4H are more favorable candidates for the formulation of targeted drug delivery. These polymers bearing the desired functional group interact with mucus through physical entanglement as well as through chemical bonds resulting in the formation of a cross-linked network. For example, urea is a well-accepted hydrogen-bonding disruptor that decreases the mucoadhesion of mucin/pectin samples. Other properties that may affect the mucoadhesive nature of the polymer include chain length, degree of hydration, degree of cross-linking, polymer concentration, charge, etc.

In the last three decades, periodontal therapy has seen significant progress in various aspects. There has been a shift from surgical treatment procedures to techniques and methods aimed at delivering the drug locally along with scaling and root planing to the affected sites by targeting the specific periodontopathic microorganisms in bringing improvements in clinical parameters of the periodontium. The adjunctive use of antimicrobial agents in addition to non-surgical therapy has been shown to provide additional benefits. Such antimicrobial therapy as an adjunct needs a reduced dosage and fewer applications and should also have high patient compliance. Various LDD systems have been introduced to overcome the disadvantages of the systemic route of antimicrobial administration. A number of LDD systems have been used in different clinical trials with different degrees of success. Upon analysis of the various clinical reports, it is seen that most of the LDD systems have resulted in significant improvement in the clinical parameters. However, the kind of improvement in the clinical parameters has not been consistent. Many of the published research on LDD systems in the literature has not evaluated the change in the microbial count of periodontopathic bacteria.

In the present research, we have used Carbopol 934 for the formulation of 2% lemongrass gel, which eventually yielded many therapeutic benefits by releasing the drug in a sustained manner. Carbopol or carbomer are high molecular weight polymers of acrylic acid cross-linked with either allyl sucrose or allyl ethers of penta erythritol. These contain 56% and 68% of carboxylic acid groups calculated on the dry bases [26]. These are used as suspending agents or viscosity-increasing agents, dry and wet binders, as well as rate-controlling agents in tablets, enzyme inhibitors of intestinal protease in peptide-containing dosage forms, etc. Carbomer is a pH-dependent polymer that stays in solution form at acidic pH but forms a low viscosity gel at alkaline pH. Carbopol offers the advantage of exhibiting excellent mucoadhesive properties in comparison with other polymers (e.g., cellulose derivatives and polyvinyl alcohol) [27]. Different mucoadhesive formulations containing carbopol have been developed and it was found that these demonstrated excellent mucoadhesive properties and release the drug in a controlled manner for a longer period of time. Tan et al. [28] developed a bioadhesive gel incorporating lidocaine using carbopol and Polyvinylpyrrolidone (PVP). The results indicated that an increase in carbopol concentration significantly increased gel compressibility, hardness, and adhesiveness, that is, the factors that affect the ease of gel removal from the container, ease of gel application onto the mucosal membrane, and gel bioadhesion, respectively. Moreover, the resulting formulation provided a sustained release as compared with the conventional dosage forms. Similar results were obtained by Bilensoy et al. [29]. They developed 5-FU containing thermosensitive, mucoadhesive gel based on carbopol 934 and pluronic F12 for the treatment of Human Papillomavirus (HPV)-induced cervical cancer. The resulting formulation demonstrated better anticancer activity at lower doses, avoiding unwanted side effects of the drug. In another study, Patel and Chavda [30] prepared amoxicillin-loaded gastroretentive microspheres using carbopol-934 providing sustained release.

In the present study, there was significant reduction in all clinical mean scores from baseline to 1st and 3rd month follow-up in both 2% lemongrass gel and 10% Doxycycline Gel groups. Similarly, there was significant reduction in all microbiological mean CFU scores from baseline to 1st and 3rd month follow-up in both 2% Lemongrass Gel and 10% Doxycycline Gel groups. The mean differences for both groups from baseline to 1st and 3rd month follow-up scores showed that there was significant reduction of GI, PPD, CAL and all three microbiological mean CFU count scores in both 2% Lemongrass Gel and 10% Doxycycline Gel groups. The mean difference was highest in PPD and CAL assessments from baseline to follow-up visits in both groups. The mean differences of microbial CFU's from baseline to 1st and 3rd month follow-up in both groups were comparable.

This is attributable to the antibacterial property of doxycycline and lemongrass. Studies have shown that the beneficial effects of doxycycline in periodontal diseases are due to the antibacterial property of doxycycline against periodontopathogens [6] and the inhibitory action of the pathologically elevated tissue degrading activities of matrix metalloproteinases (MMPs) in the inflamed gingival tissues of adult periodontitis [9]. The control group-associated results of the present study are in accordance with the findings of the previous studies [6,24].

Lemongrass has several beneficial properties that can be of use in periodontal therapy. Previous studies have demonstrated the anti-inflammatory and antimicrobial properties of lemongrass in terms of inhibition of the production of interleukine-1β (IL-1β) and IL-6 [31]. Citral, the main component of the lemongrass, is responsible for its anti-microbial property by causing extensive leakage of critical molecules and ions from the bacterial cell and permeabilization of the bacterial cytoplasmic membrane leading to their death [32] and in addition to this, the anti-inflammatory action of lemongrass is due to the blockage of the LPS-induced activation of Nuclear Factor kappa-B (NF-kB) [33,34]. The observations of the test group in the present study are in accordance with the findings of previous studies [17,35,36]. In addition to this, lemongrass oil has also been shown to decrease the volatile sulfur compounds, hence inhibiting halitosis [36].

There was no statistically significant difference in mean PI scores from baseline to the 1st and 3rd month follow-up scores in the 2% lemongrass gel group. This could be because of a lack of oral hygiene maintenance.

The results of this study suggested that 2% lemongrass gel as an LDD system is one of the nonsurgical treatment modalities in bringing improvement in clinical & microbiological parameters. However, further investigation with a larger sample size on a prolonged post-operative follow-up is required to conclusively establish the effectiveness of this gel.

5. Conclusions

It could be concluded that the local delivery of 2% lemongrass gel as an adjunct to scaling and root planing is effective and comparable to 10% doxycycline gel in the treatment of chronic periodontitis, and 2% lemongrass gel is an effective antimicrobial agent against primary periodontal pathogens, i.e., *Porphyromonas gingivalis, Actinomyces naeslundii,* and *Prevotella intermedia*. Further studies targeting other periodontal pathogens associated with specific periodontal conditions need to be conducted.

Author Contributions: Conceptualization, P.M., S.T.G., S.M., S.M.A.-Q., M.A.M., R.R.N., S.T., S.C., A.A. and V.V. methodology, P.M., S.T.G., S.M., M.A.M., R.R.N. and S.C.; software, P.M., S.T.G. and S.M.; validation, P.M., S.T.G., S.M., R.R.N., S.T. and V.V.; formal analysis, P.M., S.T.G. and S.M.; investigation, P.M., S.T.G., S.M. and S.M.A.-Q.; resources, P.M., S.T.G., S.M. and A.A.; data curation, P.M., S.T.G. and S.M.; writing—original draft preparation, P.M., S.T.G. and S.M.; writing—review and editing, P.M., S.T.G., S.M., S.M.A.-Q., M.A.M., R.R.N., S.T., S.C., A.A. and V.V.; visualization, P.M., S.T.G. and S.M.; supervision, P.M., S.T.G. and S.M.; project administration, P.M., S.T.G., S.M. and S.C.; funding acquisition, P.M., S.T.G., S.M., S.M.A.-Q., M.A.M., R.R.N., S.C., A.A. and V.V.; All authors have read and agreed to the published version of the manuscript.

Funding: The authors extend their appreciation to the Deanship of Scientific research at King Khalid University, Saudi Arabia for funding this work through Small Group Project under the grant number (RGP.1/204/43).

Institutional Review Board Statement: The study was conducted in accordance with the Declaration of Helsinki, and the protocol was approved by the Ethics Committee of Institute of dental science, Bareilly, India. "(IDS/ETHCC/14/10)." The study was registered with the Clinical Trial Registry of India (CTRI REF/2021/03/042330 AU).

Informed Consent Statement: Informed consent was obtained from all subjects involved in the study.

Data Availability Statement: Data can be made available on demand by the chief researcher for academic purposes by email.

Conflicts of Interest: The authors declare no conflict of interest. The funders had no role in the design of the study; in the collection, analyses, or interpretation of data; in the writing of the manuscript, or in the decision to publish the results.

References

1. Newman, M.G.; Takei, H.; Klokkevold, P.R.; Carranza, F.A. *Newman and Carranza's Clinical Periodontology E-Book*; Elsevier Health Sciences: IOWA City, IA, USA, 2018; pp. 50–370.
2. Salvi, G.E.; Mombelli, A.; Mayfield, L.; Rutar, A.; Suvan, J.; Garrett, S.; Lang, N.P. Local antimicrobial therapy after initial periodontal treatment. *J. Clin. Periodontol.* **2002**, *29*, 540–550. [CrossRef] [PubMed]
3. Rams, T.E.; Slots, J. Local delivery of antimicrobial agents in the periodontal pocket. *Periodontology 2000* **1996**, *10*, 139–159. [CrossRef] [PubMed]
4. Socransky, S.S.; Haffajee, A.D.; Cugini, M.A.; Smith, C.; Kent, R.L., Jr. Microbial complexes in subgingival plaque. *J. Clin. Periodontol.* **1998**, *25*, 134–144. [CrossRef] [PubMed]
5. Zambon, J.J. Periodontal diseases: Microbial factors. *Ann. Periodontol.* **1996**, *1*, 879–925. [CrossRef]
6. Walker, C.B.; Godowski, K.C.; Borden, L.; Lennon, J.; Nangó, S.; Stone, C.; Garrett, S. The effects of sustained release doxycycline on the anaerobic flora and antibiotic-resistant patterns in subgingival plaque and saliva. *J. Periodontol.* **2000**, *71*, 768–774. [CrossRef]
7. Soskolne, W.A. Subgingival delivery of therapeutic agents in the treatment of periodontal diseases. *Crit. Rev. Oral Biol. Med.* **1997**, *8*, 164–174. [CrossRef]
8. Slots, J.; Rosling, B.G. Suppression of the periodontopathic microflora in localized juvenile periodontitis by systemic tetracycline. *J. Clin Periodontol.* **1983**, *10*, 465–486. [CrossRef]

9. Golub, L.M.; Sorsa, T.; Lee, H.M.; Ciancio, S.; Sorbi, D.; Ramamurthy, N.S.; Gruber, B.; Salo, T.; Konttinen, Y.T. Doxycycline inhibits neutrophil (PMN)-type matrix metalloproteinases in human adult periodontitis gingiva. *J. Clin. Periodontol.* **1995**, *22*, 100–109. [CrossRef]
10. Jorgensen, M.G.; Slots, J. Responsible use of antimicrobials in periodontics. *J. Calif. Dent. Assoc.* **2000**, *28*, 185–193.
11. Slots, J.; Pallasch, T.J. Dentists' role in halting antimicrobial resistance. *J. Dent. Res.* **1996**, *75*, 1338–1341. [CrossRef]
12. Palombo, E.A. Traditional Medicinal Plant Extracts and Natural Products with Activity against Oral Bacteria: Potential Application in the Prevention and Treatment of Oral Diseases. *Evid. Based Complement. Altern. Med.* **2011**, *2011*, 680354. [CrossRef]
13. Khan, R.; Islam, B.; Akram, M.; Shakil, S.; Ahmad, A.; Ali, S.M.; Siddiqui, M.; Khan, A.U. Antimicrobial activity of five herbal extracts against multi drug resistant (MDR) strains of bacteria and fungus of clinical origin. *Molecules* **2009**, *14*, 586–597. [CrossRef]
14. Sharma, R.; Hebbal, M.; Ankola, A.V.; Murugaboopathy, V.; Shetty, S.J. Effect of two herbal mouthwashes on gingival health of school children. *J. Tradit. Complement. Med.* **2014**, *4*, 272–278. [CrossRef]
15. Shah, G.; Shri, R.; Panchal, V.; Sharma, N.; Singh, B.; Mann, A.S. Scientific basis for the therapeutic use of Cymbopogon citratus, stapf (Lemon grass). *J. Adv. Pharm. Technol. Res.* **2011**, *2*, 3–8. [CrossRef]
16. Hammer, K.A.; Carson, C.F.; Riley, T.V. Antimicrobial activity of essential oils and other plant extracts. *J. Appl. Microbiol.* **1999**, *86*, 985–990. [CrossRef]
17. Khongkhunthian, S.; Sookkhee, S.; Okonogi, S. Antimicrobial activities against periodontopathogens of essential oil from lemon grass (Cymbopogon citratus (DC.) Stapf CMU. *J. Nat. Sci.* **2009**, *8*, 11–21.
18. Susanto, S.A.; Oktavianti, T.A.; Wijaya, Y.; Wira, V.; Paramitta, V.A. Increased glutathione level in saliva of moderate gingivitis patients after lemongrass (cymbopogon citratus) essential oil gargling. *Asia Pac. Dent. Stud. J.* **2010**, *1*, 45–52.
19. Anand, K.M.; Goyal, R.; SubrayaBhat, G.; Kamath, S.; Aggarwal, M.; Meghna, A.; Bhandarkar, B.S.; Sukreeth, S. A novel anti-oxidant lemon grass oil mouthwash—A clinical trial. *Asian J. Exp. Biol. Sci.* **2011**, *2*, 482–486.
20. Patel, J.; Patel, B.; Banwait, H.; Parmar, K.; Patel, M. Formulation and Evaluation of Topical Aceclofenac Gel Using Different Gelling Agents. *Int. J. Drug Dev. Res.* **2011**, *3*, 156–164.
21. Armitage, G.C. Development of a classification system for periodontal diseases and conditions. *Ann. Periodontol.* **1999**, *4*, 1–6. [CrossRef]
22. Loe, H.; Silness, J. Periodontal disease in pregnancy. I. Prevalence and severity. *Acta Odontol. Scand.* **1963**, *21*, 533–551. [CrossRef]
23. Silness, J.; Loe, H. Periodontal disease in pregnancy. Correlation between oral hygiene and periodontal condition. *Acta Odontol. Scand.* **1964**, *22*, 121–135. [CrossRef]
24. Shaddox, L.M.; Andia, D.C.; Casati, M.Z.; Nociti, F.H., Jr.; Sallum, E.A.; Gollwitzer, J.; Walker, C.B. Microbiologic changes following administration of locally delivered doxycycline in smokers: A 15-month follow-up. *J. Periodontol.* **2007**, *78*, 2143–2149. [CrossRef]
25. Gu, J.M.; Robinson, J.R.; Leung, S.H. Binding of acrylic polymers to mucin/epithelial surfaces: Structure-property relationships. *Crit. Rev. Ther. Drug Carr. Syst.* **1988**, *5*, 21–67.
26. Barry, B.W.; Meyer, M.C. The rheological properties of carbopol gels I. Continuous shear and creep properties of carbopol gels. *Int. J. Pharm.* **1979**, *2*, 1–25. [CrossRef]
27. Davies, N.M.; Farr, S.J.; Hadgraft, J.; Kellaway, I.W. Evaluation of mucoadhesive polymers in ocular drug delivery. II. Polymer-coated vesicles. *Pharm. Res.* **1992**, *9*, 1137–1144. [CrossRef]
28. Tan, Y.T.; Peh, K.K.; Hanbali, A.O. Effect of Carbopol and polyvinylpyrrolidone on the mechanical, rheological, and release properties of bioadhesive polyethylene glycol gels. *AAPS PharmSciTech* **2000**, *1*, 69–78. [CrossRef]
29. Bilensoy, E.; Sen, M.; Dogan, A.L.; Calis, S. Thermosensitive mucoadhesive gel formulation loaded with 5-Fu: Cyclodextrin complex for HPV-induced cervical cancer. *J. Incl. Phenom. Macrocycl. Chem.* **2007**, *57*, 363–370. [CrossRef]
30. Patel, J.K.; Chavda, J.R. Formulation and evaluation of stomachspecific amoxicillin-loaded carbopol-934P mucoadhesive microspheres for anti-Helicobacter pylori therapy. *J. Microencapsul.* **2009**, *2*, 6365–6376.
31. Sforcin, J.; Amaral, J.; Fernandes, A., Jr.; Sousa, J.; Bastos, J. Lemongrass effects on IL-1β and IL-6 production by macrophages. *Nat. Prod. Res.* **2009**, *23*, 1151–1159. [CrossRef]
32. Sikkema, J.; de Bont, J.A.; Poolman, B. Interactions of cyclic hydrocarbons with biological membranes. *J. Biol. Chem.* **1994**, *269*, 8022–8028. [CrossRef]
33. Lee, H.J.; Jeong, H.S.; Kim, D.J.; Noh, Y.H.; Yuk, D.Y.; Hong, J. Inhibitory effect of citral on NO production by suppression of iNOS expression and NF-κB activation in RAW264. 7 cells. *Arch. Pharm. Res.* **2008**, *31*, 342–349. [CrossRef] [PubMed]
34. Bachiega, T.F.; Sforcin, J.M. Lemongrass and citral effect on cytokines production by murine macrophages. *J. Ethnopharmacol.* **2011**, *137*, 909–913. [CrossRef] [PubMed]
35. Warad, S.B.; Kolar, S.S.; Kalburgi, V.; Kalburgi, N.B. Lemongrass essential oil gel as a local drug delivery agent for the treatment of periodontitis. *Anc. Sci. Life* **2013**, *32*, 205–211.
36. Satthanakul, P.; Taweechaisupapong, S.; Paphangkorakit, J.; Pesee, M.; Timabut, P.; Khunkitti, W. Antimicrobial effect of lemongrass oil against oral malodour micro-organisms and the pilot study of safety and efficacy of lemongrass mouthrinse on oral malodour. *J. Appl. Microbiol.* **2015**, *118*, 11–17. [CrossRef]

polymers

MDPI

Article

Phase-Inverted Copolymer Membrane for the Enhancement of Textile Supercapacitors

Sheng Yong [1,*], Nicholas Hillier [2] and Stephen Paul Beeby [1]

[1] Smart Electronic Materials & System Research Group, School of Electronics and Computer Science, University of Southampton, Southampton SO17 1BJ, UK
[2] Energy Storage and Its Applications Centre of Doctoral Training, University of Southampton, Southampton SO17 1BJ, UK
* Correspondence: s.yong@soton.ac.uk; Tel.: +44-2380593234

Abstract: This paper presents a universal fabrication process for single-layer textile supercapacitors, independent of textile properties such as weave pattern, thickness and material. To achieve this, an engineered copolymer membrane was fabricated within these textiles with an automated screen printing, phase inversion and vacuum curing process. This membrane, together with the textile yarns, acts as a porous, flexible and mechanically durable separator. This process was applied to four textiles, including polyester, two polyester-cottons and silk. Carbon-based electrodes were subsequently deposited onto both sides of the textile to form the textile supercapacitors. These supercapacitors achieved a range of areal capacitances between 3.12 and 38.2 mF·cm^{-2}, with energy densities between 0.279 and 0.681 mWh·cm^{-3} with average power densities of between 0.334 and 0.32 W·cm^{-3}. This novel membrane facilitates the use of thinner textiles for single-layered textile supercapacitors without significantly sacrificing electrochemical performance and will enable future high energy density textile energy storage, from supercapacitors to batteries.

Keywords: supercapacitor; e-textile; phase inversion membrane; energy storage device; polymer separator; carbon electrode

1. Introduction

The integration of electronics within textiles (e-textiles) has seen a plethora of applications unlocked for medical [1], defence [2] and consumer [3] wearable technologies. Increasingly though, the power management of these technologies has become a bottleneck for further designs. This requirement for power bridging (analogous to increasing the penetration of renewable energy onto electrical networks) or power averaging has seen a stark increase in research interest in textile supercapacitors (TSCs) and batteries. This is highlighted by an increase in publications of two orders of magnitude over the previous decade [4]. TSCs have been far more widely researched than other textile energy storage systems due to their increased energy and power density when compared to traditional capacitors [5] and their longer cycle lifetime and reliability when compared to batteries [6]. Similar to a traditional capacitor, a supercapacitor comprises two electrodes, an electrolyte and an electrically insulating but ionically conductive separator. To date, electrode materials for TSCs have seen extensive research with advances in conductivity [7], porosity [8], mechanical and chemical stability [9] and even functionalisation [10]. The electrolyte likewise has seen significant research with a lot of the focus on aqueous gel-based formulations (H_2SO_4 [11], H_3PO_4 [12], KOH [13]), but also some exciting explorations into the use of nanoparticle fillers [14]. Despite this, the separator layer has seen little advancement, with it often being realised simply as an extra layer of separator paper sandwiched between textile electrodes [15] or as an additional film layer [16] in a sandwich-like configuration. These solutions impair the wearable technology, with asymmetric stress within the layers induced by the movement of the wearer and impracticalities of having extra layers within an encapsulated textile system. Sheng et al. [17] achieved a single-layer TSC using the fabric itself as the separator, impregnating a central region with a gel electrolyte. However, the cotton substrate had to be 500 μm thick and required a very dense weave. Though this thickness allowed for the separation of the carbon electrodes, later work by the group [18] showed that for multilayer configurations, a thinner fabric achieved much higher capacitances. Clearly, there is a requirement to integrate a separator layer into thinner textiles to enable thinner but more energy-dense TSCs. One method to achieve this is to fill the gaps between the textile fibres/yarns with a porous membrane before the deposition of the electrodes and the impregnation with the electrolyte.

The design of separator membranes is important for all kinds of electrochemical energy storage, from supercapacitors to fuel cells. Important characteristics include chemical stability, ionic conductivity, flexibility, electrolyte update and porosity. Different crystalline and amorphous materials have been evaluated for the fabrication of discrete separator membranes. These include cellulose [19], aluminium oxide nanowire with polyvinyl butyral (PVB) [20], polyacrylonitrile (PAN) [21], polyvinylidene fluoride (PVDF) [22], polyvinyl alcohol (PVA)-tetraethyl orthosilicate [23], poly arylene ether sulfone/polyethylene oxide (PEO) [24] and PVDF-poly methyl methacrylate (PMMA) composite [25]. These materials have a high degree of processability and can be readily fabricated into a porous separator membrane via electrospinning [26] or phase inversion [27]. Phase inversion is a process of immersing a liquid-polymer solution into an anti-solvent solution which removes the original solvent and leaves a porous, solid polymer membrane. However, with the primary research focusing on traditional battery and fuel cell technologies, many of these separators are not suitable for e-textile applications due to poor adhesion to, or integration within, the textile substrate. For example, PAN-based separator membranes have high electrolyte uptake but also have electrolyte leakage issues, PVDF-based separator membranes have high ionic conductivity and high chemical and mechanical resistivity, but their hydrophobicity limits the electrolyte uptake and retention [28]. PMMA-based separator membranes have high porosity and good compatibility with electrolytes due to the polymer's amorphous nature, but their poor mechanical strength limits their use in energy storage devices [29]. This issue has been improved by mixing and cross-linking different amorphous polymers together with (or without) metal oxide nano/microwires [30]. A separator membrane fabricated in this manner can achieve high porosity, good compatibility with electrolytes and improved mechanical durability suitable for energy storage device design. However, adding these materials to textiles requires additional fabrication processes such as spinning, lamination or embroidery and presents a sub-optimal textile separator design [31]. Therefore, it is essential to develop materials and methodologies that can integrate a membrane layer within a standard textile, utilizing traditional textile fabrication processes. One solution is to introduce polymers such as ethylene-vinyl acetate (EVA) into the separator membrane material matrix to help it fill the space within and adhere to the textile substrate forming a textile and polymer-based separator membrane. EVA is a colourless, flexible and waterproof copolymer of ethylene and vinyl acetate and is a well-known adhesive material for textiles, paper, and other porous materials. The combination of EVA and PMMA creates a mechanically flexible adhesive for optical fibre bonding [32]. The addition of EVA changes the fracture mechanism of the PMMA polymer matrix from crazing to yielding, improving its flexibility [33]. Additionally, numerous fibrils and voids/pores form in the solution-processed EVA/PMMA polymer matrix. These voids also can be used as tunnels and pores for the electrolyte flow, making them suitable as a porous material for separator design.

In this paper, we report a novel approach to the fabrication of carbon-based TSCs, making use of a membrane-impregnated textile substrate. The membrane layer of the proposed supercapacitor was produced via screen printing, followed by a phase inversion process and was composed of a copolymer blend of EVA and PMMA. The effect of polymer ratios on the performance of the membrane is investigated and characterised through microscopy and electrochemical techniques. The reported methodology was shown to be universal across four standard textiles (one polyester, two different polyester-cotton and one silk substrate), with all samples showing promising electrochemical performances and a reduction in process complexity during the electrode deposition stage.

2. Materials and Methods

2.1. Material

Activated carbon YP-80F was obtained from Kuraray Chemical (Tokyo, Japan), EVA beads (vinyl acetate 25 wt. %), PMMA powder (Mw = 120,000 g/mol), carbon nano-powder additive (<100 nm particle size), 1,2,4-trichlorobenzene, ethanol, dimethyl sulfoxide and Isoamyl acetate were acquired from Sigma-Aldrich (Gillingham, UK). Polyester (denoted as P), two polyester cotton textiles (denoted as PC and PC (thin) and silk (denoted as S) were obtained from Klopman (Frosinone, Italy). The textiles were all plain weaves, with further characteristics presented in Table 1.

Table 1. Characteristics of the textile substrates used throughout this study.

	Polyester-Cotton/Thin (PC Thin)	Polyester-Cotton (PC)	Silk (S)	Polyester (P)
Fibre types	Cotton and Polyester	Cotton and Polyester	Silk	Polyester
Fibre diameter (μm)	12 (Polyester) 15 (Cotton)	12 (Polyester) 15 (Cotton)	~8.3	~20
Ends per inch	16.5	16.5	19.6	13.5
Picks per inch	9.05	9.05	15.7	7.4
Textile original weight ($mg{\cdot}cm^{-2}$)	16.3	27.1	4.45	23.4
Textile thickness (μm)	150	250	50	244

2.2. Preparation of Textile with a Polymer Membrane and Bending Experiment

Screen printing and phase inversion processes were applied to fabricate the EVA-PMMA copolymer membrane in the cotton textile. The copolymer solution was prepared by mixing EVA beads (vinyl acetate 25 wt. %) with PMMA powder ratios of 1:9, 3:7 and 5:5 by weight, and the textiles with these polymer membranes were denoted as PC1:9, PC3:7 and PC5:5. In total, 2 g polymer blends were dissolved in a solvent mixture of 2.5 mL of 1,2,4-Trichlorobenzene and 2.5 mL of Isoamyl acetate. The polymer blends were completely dissolved in the solvent mixture via heating at 60 °C with a magnetic stirring bar rotated at 600 rpm until the solution became clear. The copolymer solution was then screen-printed twice onto the top of the textile with a screen mesh thickness of 40 μm, printing gap of 0.8 mm, printing pressure of 5.5 kg and speed of 10 $mm{\cdot}s^{-1}$. After printing, the copolymer-coated cotton was treated with a phase inversion process by immersing the samples in ethanol within a sonication bath for 15 min. The samples were then cured under a vacuum for 2 h at room temperature. After the curing process, some of the copolymer was immersed into the textile and the remaining copolymer formed a membrane layer of about 100 μm on top of the textile.

A group of polyester-cotton textiles with different membranes were bent around a mandrel shown in the Supplementary Materials (Figure S2). These devices were bent around a 3.2 mm diameter mandrel for 500 or 1000 cycles at a cyclical rate of 2 cycles per second.

2.3. Fabrication of Supercapacitor Electrodes

The carbon solution was prepared with activated carbon (YP-80F is from Kuraray Chemical), conductive carbon additive and EVA beads (vinyl acetate 25 wt. %) dissolved in a 1,2,4-Trichlorobenzene solvent. The ratio of carbon to binder was 85:15 by weight [34]. The surface of the four textiles was first shaved to reduce pilosity, and the carbon solution was spray coated through a mask on both sides of the textile substrates. The spray nozzle (diameter of 0.3 mm) was positioned 15 cm away from the fabric, and air pressure of 25 psi (1.72 bars) was used for the spraying. Each side of the textile sample was sprayed for 5 s.

The mask was removed, and the textile substrate was dried in a fan oven at 120 °C for 10 min. The amount of carbon on the polyester and two polyester-cotton-based devices was 3.56 mg·cm^{-2}. For the silk device, the weight of carbon was 1.78 mg·cm^{-2} (maximum possible loading).

2.4. Textile Supercapacitor Assembly and Testing

The textile devices were cut into round pieces with a diameter of 1 cm. The organic electrolyte was developed by mixing 1.08 g of Tetraethylammonium tetrafluoroborate salt in 5 mL dimethyl sulfoxide, as demonstrated in a previous paper [35]. The organic mixture was under constant stirring until the mixture became clear. The textile electrode was immersed in the organic electrolyte under vacuum at ~25 mbar for 20 min. For testing, the single-layer textile supercapacitor was placed under compression using spring-loaded grade 303 stainless steel current collectors housed within a Swagelok PFA tube fitting (Swagelok, Solon, OH, USA).

2.5. Characterisation and Electrochemical Testing of Flexible Textile Supercapacitor

The device's surface profile and cross-sectional structure were obtained via scanning electron microscopy (SEM) using a ZES EVO microscope at an operating voltage of 5 kV, 1.68 A with different magnifications. The porosity of the PC textile and PC3:7 textile was obtained by Equation (1) and the distilled water immersion method [36]. Both samples were immersed in the distilled water under vacuum at ~25 mbar for 20 min and then taken out. The excess distilled water on the surface was removed with filter paper, and the content of absorbed distilled water was calculated. The density of the PC textile and PC3:7 were 1.13 and 1.45 g·cm^{-3}, respectively.

$$porosity\ \% = \frac{M_d/\rho_d}{\left(M_d/\rho_d + M_0/\rho_0\right)} \times 100\% \quad (1)$$

where M_0 and ρ_0 are the weight and density of the PC textile or PC3:7 before vacuum impregnation. M_d and ρ_d are the weight and density of the distilled water absorbed into the PC textile and PC3:7 textile.

The electrolyte uptake of the PC textile and PC3:7 textile with the proposed membrane were measured via a similar methodology to the porosity test but replaced the distilled water with the proposed electrolyte. The amounts of the electrolyte uptake after the vacuum impregnation process and during the ageing test were determined by Equation (2):

$$uptake\ \% = M_e/M_0 \times 100\% \quad (2)$$

where M_0 is the weight of PC textile or PC3:7 textile with a membrane before vacuum impregnation and M_e is the weight of PC textile or PC3:7 textile with membrane after vacuum impregnation with the electrolyte and after 2, 4, 6, 21 and 24 h left in room temperature at 23 °C. The electrochemical performance of the textile supercapacitor was evaluated using an Autolab PGSTAT101 (Metrohm Autolab, Utrecht, The Netherlands). Electrochemical impedance spectroscopy measurements were performed on a Solartron 1470E Cell Test System (Solartron Analytical, Farnborough, UK) from 250 kHz to 0.1 Hz with an amplitude of 10 mV. The calculations for capacitance, energy and power were based on the GC test result that follows Equations (3)–(5).

$$C = I \times \left(dV/dt\right)^{-1} \quad (3)$$

$$E = 0.5 \times C \times V_{peak}^2 \quad (4)$$

$$P = E/t \quad (5)$$

where C is the capacitance, I is the cycling current in the GC test, dV/dt is the rate of change of voltage between 1.44 V (80% V_{peak}) and 0.72 V (40% V_{peak}) in the GC test, E is the energy stored in the device, V_{peak} is 1.8 V, P is the average power and t is the time taken for the supercapacitor to fully discharged.

3. Results and Discussion

3.1. SEM Imaging

In this work, polyester (denoted as P), two polyester cotton textiles (denoted as PC and PC (thin)) and silk (denoted as S) were used as textile substrates. The copolymer solution was prepared by mixing EVA beads with PMMA powder at ratios (EVA:PMMA) of 1:9, 3:7 and 5:5 by weight, and the textiles with these polymer membranes are denoted as P/PC/PC (thin)/S 1:9, 3:7 and 5:5.

In the EVA-PMMA membrane design, the PMMA contributes toward the porosity of the separator membrane while the EVA enhances the flexibility and adhesion. Higher percentage weights of EVA within the copolymer reduce the shrinkage of the membrane during the phase inversion and thus increase the total coverage of the membrane. However, once the EVA content reaches a critical limit, the membrane loses its porosity and the electrolyte flow through the membrane is restricted. This balancing of the ratio of EVA to PMMA was initially explored across three ratios with the resulting membranes shown in Figure 1a–c, which shows a comparison between samples PC 5:5, PC 3:7 and PC 1:9. The PC 1:9 sample in Figure 1a has less of the membrane layer present in the textile. This is due to the lower mass ratio of EVA within the membrane polymer lattices, resulting in more membrane loss, through shrinkage, during the phase inversion. This shrinkage is uncontrollable and can result in detrimental macro porosity and material loss at the edges of the device. In comparison with the other samples, this membrane is also unlikely to be able to block the electrode material upon deposition, leading to an increased risk of short circuits. The SEM photograph in Figure 1b shows the cross-section view of the sample PC 3:7, where the membrane forms a layer-by-layer structure across the gap between polyester-cotton yarns. As shown in Figure 1c, the PC 5.5 membrane with the lowest percentage of PMMA and the highest percentage of EVA fills the entire area between textile yarns. This will prevent short circuits but will also block the electrolyte flow through the device, reducing the capacitance and increasing the resistance. With the 3:7 ratio demonstrating superior bulk properties, further investigation of this polymer ratio was undertaken. Figure 1d show that the printing and phase inversion process created a porous EVA-PMMA copolymer membrane that adheres to the yarns and fibres of the textile. The membrane itself consisted of polymer lattices with micropores of around 2 μm in diameters (Figure 1e). These micropores are produced uniformly during the phase inversion process and allow for the free movement of the ions within the electrolyte.

Having formed the porous copolymer membrane separator, the next stage in the fabrication process is to deposit the carbon electrodes. For comparison purposes, carbon electrodes were spray coated onto the textiles with and without the membrane in place. Figure 2 show SEM photos of polyester-cotton (PC) and silk (S) textile before and after the spray deposition of the carbon electrode. The SEM photos of the other two textiles (P and PC (thin)) are very similar in appearance to the SEM photos of the PC textile, and they are shown in the Supplementary Materials (Figure S1). The SEM photos in Figure 2a,b show that both the PC and S textiles contain gaps between each fibre. This structure leads to the electrodes contacting each other during the electrode spray coating process, which causes short circuits. This is demonstrated in Figure 2e,f where the SEM photos for spray-coated textiles (PC and S) without the membrane show the carbon layer travelling through the textile structure. This causes carbon material to potentially penetrate through to the other side of the textile causing short-circuits. In comparison, the SEM photos in Figure 2c,d,g,h show that a polymer membrane with an EVA:PMMA ratio of 3:7 overcomes the membrane shrinkage issue observed with the EVA:PMMA ratio of 1:9, creating a homogeneous polymer layer. It not only acts as an electrical separator in the textile supercapacitors but also fills the gap between textile yarns reducing the surface roughness of the textile and blocking any carbon permeation through the substrate. This enhances the carbon deposition and forms a continuous and uniform distributed conductive electrode that leads to a lower electrode resistance and higher power density.

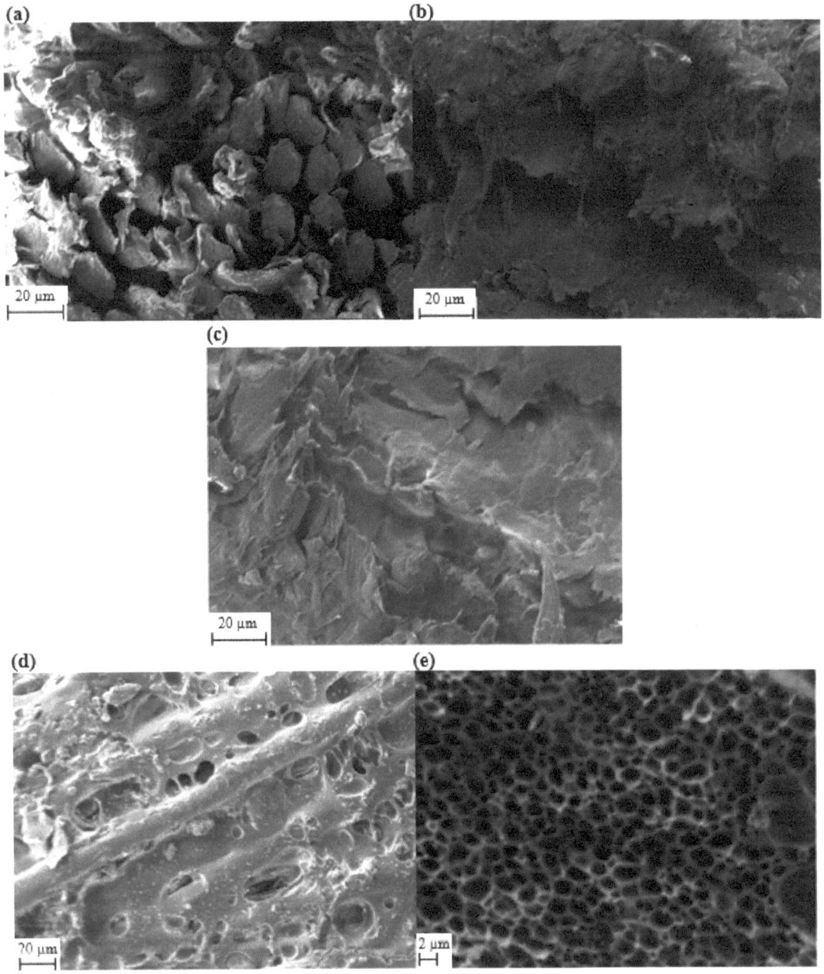

Figure 1. (**a**) Cross-section view SEM photo of sample PC 1:9, (**b**) sample PC 3:7, (**c**) sample PC 5:5, (**d**) plain view SEM photo of sample PC 3:7 and (**e**) high magnification plain view SEM photo of sample PC 3:7.

Figure 2. SEM photos showing plan views of (**a**) PC textile, (**b**) S textile, (**c**) PC textile with membrane and carbon coating, (**d**) S textile with membrane and carbon coating, (**e**) PC textile with carbon coating without membrane, (**f**) S textile with carbon coating without membrane, (**g**) PC textile with membrane and carbon coating (high magnification) and (**h**) S textile with membrane and carbon coating (high magnification).

3.2. Porosity, Electrolyte Uptake and Ageing Test

Based on the distilled water immersion methodology [33], the porosity of the PC and PC 3:7 samples was evaluated. The PC 3:7 textile demonstrated a porosity of 24.9 ± 6%, which was 6.9% lower than the PC textile only (31.8 ± 4%). The results from the electrolyte uptake and ageing test are shown in Figure 3a. After wetting the samples with an electrolyte containing tetraethylammonium tetrafluoroborate salt and dimethyl sulfoxide (the same electrolyte from the electrochemical testing), the PC 3:7 textile maintained a higher (or comparable when accounting for the uncertainty) volume of electrolyte ($51.1 \pm 8\%$ mL·cm^{-2}) than the PC textile ($47.8 \pm 10\%$ mL·cm^{-2}). The samples were then stored in air at 23 °C for 24 h to assess the electrolyte retention. As can be seen, the PC 3:7 textile exhibited similar or better retention compared to the textile alone throughout ageing. The engineered copolymer membrane, containing EVA and PMMA polymers, fills the gaps between the PC textile yarns and replaces them with much smaller-sized air pores (with this observation agreeing with the SEM results in Section 2.1) and is the reason for the lower porosity. However, despite the lower porosity, the membrane-infused textile was able to absorb a greater amount of electrolyte and retained it better/or equal to the textile. In fact, after 21 h, the retention of the electrolyte was significantly better, suggesting that with further refinement, the membrane could be engineered to retain electrolytes far in excess of the textile alone. The Nyquist plot results of the PC 3:7 textile and PC textile are shown in Figure 3b, with the tests performed over the frequency range of 0.1 to 250 kHz at a 10 mV voltage amplitude. The equivalent series resistance of the PC 3:7 textile ($34.6 \pm 18.7\%$ Ω or $0.956 \pm 16.4\%$ mS·cm^{-1}) is more resistive than the PC textile ($25.4 \pm 15.6\%$ Ω or $1.30 \pm 17.5\%$ mS·cm^{-1}) (characterised close to 250 kHz) at the crossing point of the real axis. Though higher, the resistance values demonstrate that the membrane does not dramatically influence the equivalent series resistance of textile and thus does not diminish the electrochemical behaviour of the PC textile.

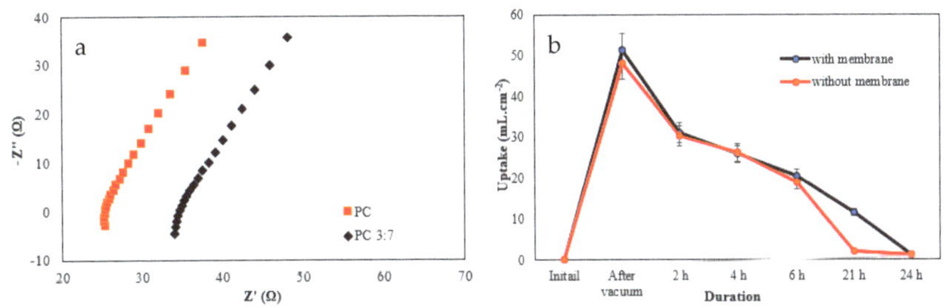

Figure 3. (**a**) Nyquist plots of PC 3:7 textile and PC textile and (**b**) electrolyte uptake and ageing test of PC 3:7 textile and PC textile.

3.3. Electrochemical Results

Each device has a surface area of 0.785 cm^2 with a thickness of approximately 244, 150, 250 and 50 μm for the P, PC (thin), PC and S textiles, respectively. The following results were averaged from five device tests. Cyclic voltammetry (CV) tests were performed at different scan rates between $+/-1.8$ V at 150 mV·s^{-1}, and Galvanostatic cycling (GC) tests were performed between 0 and 1.8 V at a cycling current of 1 mA·cm^{-2}.

The CV and GC test results for the single-layer textile supercapacitors based on the impregnated textile substrates (P, PC (thin), PC and S) with an EVA:PMMA ratio of 3:7 are shown in Figure 4a,b, respectively. The single-layer textile supercapacitors with polyester textile (type P), polyester-cotton textile (type PC) and thinner polyester-cotton textile (type PC thin) achieved areal capacitances of 29.2 mF·cm^{-2}, 35.6 mF·cm^{-2} and 27.6 mF·cm^{-2}, respectively. The single-layer silk textile supercapacitor was shown to work without short circuits and charged to $+/-1.8$ V; however, it demonstrated a reduced area capacitance of 3.12 mF·cm^{-2}. The incorporation of the copolymer membrane as a separator enables supercapacitors to be reliably fabricated in a single layer of silk which cannot be achieved without the membrane in place. However, the nature of the weave with the larger gaps means the resulting capacitor properties are the lowest of the textiles tested in this study.

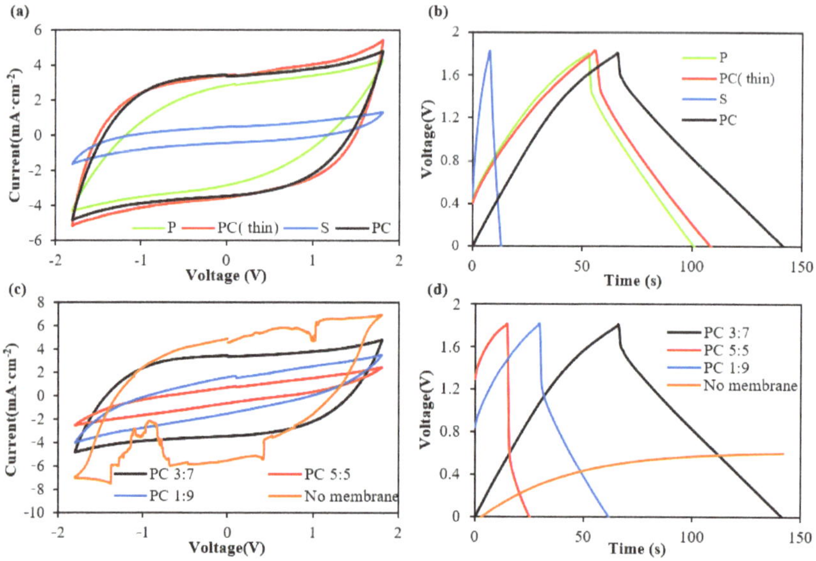

Figure 4. (**a**) CV and (**b**) GC result for single-layer textile supercapacitor based on the different textiles with an EVA:PMMA ratio of 3:7. (**c**) CV and (**d**) GC result for single-layer textile supercapacitor based on PC 1:9, PC 3:7, PC 5:5 and PC without membrane.

With the PC textile exhibiting the best electrochemical performance, this textile was further investigated to determine supercapacitor performance with the other membrane ratios. The CV and GC results for the different membrane ratios are shown in Figure 4a,b. These results indicate the single-layer textile supercapacitor based on PC 3:7 achieved a higher capacitance and a lower internal resistance of 320.4 $\Omega \cdot cm^{-2}$ (obtained from the voltage drop in GC results) compared to the devices made from the membrane-infused textiles PC 1:9 (925.7 $\Omega \cdot cm^{-2}$) and PC 5:5 (1904.1 $\Omega \cdot cm^{-2}$). Four out of five devices with the PC1:9 membrane textile short-circuited upon testing. The only successful device demonstrated an areal capacitance of 24 $mF \cdot cm^{-2}$. From the SEM analysis of the 1:9 polymer ratio, this poor reliability is attributed to over shrinkage of the membrane, leading to contact between the electrodes. All five supercapacitors with textile membrane PC5:5 successively passed the CV and GC testing but achieved a much lower capacitance of (11.9 $mF \cdot cm^{-2}$) than the textile membrane PC 1:9 and PC 3:7 supercapacitors (Figure 4c,d). This was because the higher loading of EVA in the membrane blocked the electrolyte flow between carbon electrodes (as theorised from the SEM analysis), reducing its capacitance and increasing its resistance. Whilst previous work demonstrated successful single textile layer supercapacitors fabricated in cotton [17], devices replicating this structure in PC without the membrane effectively failed and could only be charged to 0.6 V under GC cycling (Figure 4d). This demonstrates that single-layer textile supercapacitors cannot be fabricated with this kind of textile without the introduction of a membrane layer, highlighting the significance of this membrane for low-thickness textile supercapacitors.

With the introduction of a secondary layer within the textile, the mechanical stability of the TSC could be impaired. As such, the best performing device (PC 3:7) underwent a series of cyclical bending tests around a 3.2 mm diameter mandrel, with the electrochemical performance being monitored throughout. Figure 5a,b show the CV and GC test results of this single-layer textile supercapacitor based on PC3:7 before bending and after 500 and 100 bending cycles. The CV results in Figure 5a indicate that the supercapacitor's performance was not negatively influenced by the mechanical bending. As shown in Figure 5b, the discharge time of the device increased slightly, showing a capacitance increase after bending from 35.6 mF·cm^{-2} to 36.2 mF·cm^{-2} after 500 bending cycles and 37.3 mF·cm^{-2} after 1000 bending cycles. The initial IR voltage drop at the beginning of the discharge stage in the GC testing reduced significantly from 0.346 V before bending to 0.285 V after 500 bending cycles and 0.192 V after 1000 bending cycles. This change led to lower device resistance and hence improved maximum power density from 0.253 W·cm^{-3} to 0.448 W·cm^{-3} after 1000 bending cycles. The improvement in supercapacitor properties after bending may be due to the membrane relaxing and becoming fractionally more porous, but any change was too small to be observed in the SEM analysis.

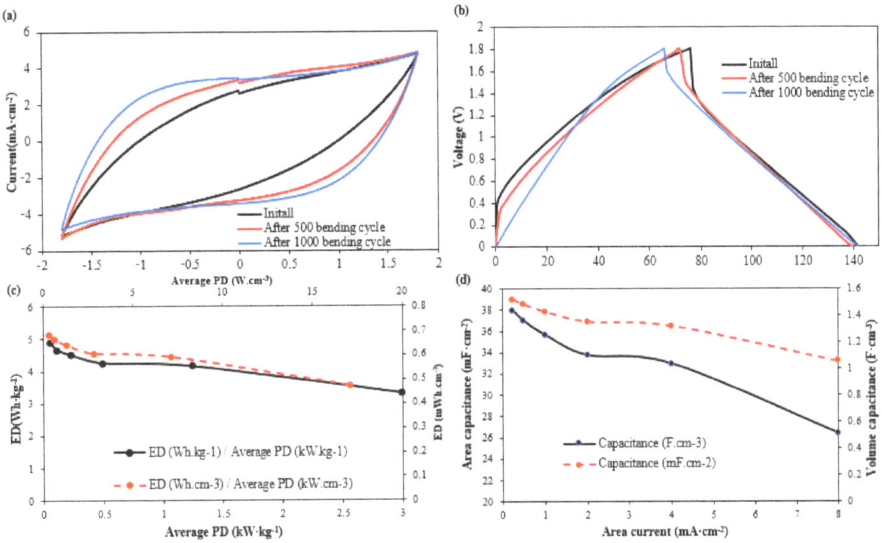

Figure 5. (**a**) CV and (**b**) GC result for single-layer textile supercapacitor based on PC3:7 before and after bending. (**c**) Energy density (ED) and average power density (PD) Ragone plot, (**d**) capacitance variation of single-layer textile supercapacitor based on PC 3:7 and extract from GC test with cycling currents are 0.25, 0.5, 1, 2, 4, 8 mA·cm^{-2}. All results are based on device thickness of 250 µm and carbon material loading of 3.56 mg·cm^{-2}.

From the results presented in Figure 5c,d, the maximum results for the single-layer textile supercapacitor based on the PC 3:7 textile (tested at a cycling current of 0.25 mA·cm^{-2}) are an areal capacitance of 38.2 mF·cm^{-2} or 1.51 F·cm^{-3} (based a device thickness of 250 µm). The maximum energy density and average power density obtained at a cycling current of 0.25 mA·cm^{-2} were 0.68 mWh·cm^{-3}/4.88 Wh·kg^{-1} and 0.32 W·cm^{-3}/0.057 kW·kg^{-1}, respectively. An increase in the cycling current to 8 mA·cm^{-2} results in increases in average power density to 17.1 W·cm^{-3} or 3 kW·kg^{-1}; however, the maximum energy density reduces to 0.47 mWh·cm^{-3}/3.32 Wh·kg^{-1} and the areal capacitance reduces to 26.4 mF·cm^{-2} or 1.06 F·cm^{-3}.

4. Conclusions

This paper presented a novel fabrication process for supercapacitors within a single textile layer, regardless of the type, thickness and structure of the fabric. This was delivered through the implementation of a phase-inverted copolymer membrane layer integrated directly within the textile. This membrane layer acts as the electrically insulating layer as well as reducing the permeation of carbon through the substrate during production. This enabled the use of ultrathin textiles as a single-layer substrate for the first time. Three polymer formulations were investigated and shown to produce large variations in both bulk material properties and electrochemical results. The membrane formed from the

3:7 ratio of EVA:PMMA was shown to have the best capacitive, mechanical and physical properties. The lower EVA content (1:9) showed significant shrinkage during the phase inversion and allowed carbon to penetrate the membrane resulting in short circuits. The higher EVA content (5:5 ratio) showed a significantly lower porosity. Electrochemically, the carbon-based TSC produced from the PC3:7 substrate demonstrated the highest areal capacitance, energy and power densities (38.2 mF·cm^2, 1.51 F·cm^3, 0.68 mWh·cm^{-3} with an average power density of 0.32 W·cm^{-3}). The membrane textile was also shown to be stable under mechanical bending of up to 1000 cycles, indicating a good level of mechanical robustness. Future work will include improving the homogeneity of the membrane and its incorporation into textile-based primary and secondary batteries.

Supplementary Materials: The following supporting information can be downloaded at: https://www.mdpi.com/article/10.3390/polym14163399/s1, Figure S1: SEM photos showing plan views of (**a**) PC 150 textile, (**b**) P textile, (**c**) PC 150 textile with membrane and carbon coating, (**d**) P textile with membrane and carbon coating, (**e**) PC 150 textile with carbon coating without membrane, (**f**) P textile with carbon coating without membrane, (**g**) PC 150 textile with membrane and carbon coating (high magnification), (**h**) P textile with membrane and carbon coating (high magnification); Figure S2: (**a**) Bending test set up, (**b**) Mandrel used in bending test.

Author Contributions: Conceptualisation, S.Y.; methodology, S.Y.; validation, S.Y.; formal analysis, S.Y. and N.H.; investigation, S.Y.; resources, S.P.B.; data curation, S.Y.; writing—original draft preparation, S.Y. and N.H.; writing—review and editing, N.H. and S.P.B.; visualisation, S.Y. and N.H.; supervision, S.P.B.; project administration, S.P.B.; funding acquisition, S.P.B. All authors have read and agreed to the published version of the manuscript.

Funding: This research was funded and supported by United State Army with Contract No: W911NF2010324 and EPSRC grants EP/P010164/1 and EP/L016818/1. The work of Steve Beeby was supported by the Royal Academy of Engineering under the Chairs in Emerging Technologies Scheme.

Data Availability Statement: The data presented in this study are available on request from the authors.

Conflicts of Interest: The authors declare no competing financial interest.

References

1. Langenhove, L.V. *Advances in Smart Medical Textiles: Treatments and Health Monitoring*, 1st ed.; Elsevier: Waltham, MA, USA, 2015.
2. Dias, T. *Electronic Textiles: Smart Fabrics and Wearable Technology*; Elsevier: Boston, MA, USA, 2015.
3. Fernandez-Carames, T.M.; Fraga-Lamas, P. Towards the Internet of Smart Clothing: A Review on IoT Wearables and Garments for Creating Intelligent Connected E-Textiles. *Electronics* **2018**, *7*, 405. [CrossRef]
4. Hillier, N.; Yong, S.; Beeby, S. The good, the bad and the porous: A review of carbonaceous materials for flexible supercapacitor applications. *Energy Rep.* **2020**, *6*, 148–156. [CrossRef]
5. Chen, H.S.; Cong, T.N.; Yang, W.; Tan, C.Q.; Li, Y.L.; Ding, Y.L. Progress in electrical energy storage system: A critical review. *Prog. Nat. Sci. Mater.* **2009**, *19*, 291–312. [CrossRef]
6. Lv, T.; Liu, M.X.; Zhu, D.Z.; Gan, L.H.; Chen, T. Nanocarbon-Based Materials for Flexible All-Solid-State Supercapacitors. *Adv. Mater.* **2018**, *30*, 1705489. [CrossRef]
7. Jin, L.N.; Shao, F.; Jin, C.; Zhang, J.N.; Liu, P.; Guo, M.X.; Bian, S.W. High-performance textile supercapacitor electrode materials enhanced with three-dimensional carbon nanotubes/graphene conductive network and in situ polymerized polyaniline. *Electrochim. Acta* **2017**, *249*, 387–394. [CrossRef]
8. Zhou, G.; Kim, N.R.; Chun, S.E.; Lee, W.; Um, M.K.; Chou, T.W.; Islam, M.F.; Byun, J.H.; Oh, Y. Highly porous and easy shapeable poly-dopamine derived graphene-coated single walled carbon nanotube aerogels for stretchable wire-type supercapacitors. *Carbon* **2018**, *130*, 137–144. [CrossRef]
9. Wang, Y.M.; Wang, X.; Li, X.F.; Li, X.L.; Liu, Y.; Bai, Y.; Xiao, H.H.; Yuan, G.H. A High-Performance, Tailorable, Wearable, and Foldable Solid-State Supercapacitor Enabled by Arranging Pseudocapacitive Groups and MXene Flakes on Textile Electrode Surface. *Adv. Funct. Mater.* **2021**, *31*, 2008185. [CrossRef]
10. Barczak, M.; Bandosz, T.J. Evaluation of nitrogen- and sulfur-doped porous carbon textiles as electrode materials for flexible supercapacitors. *Electrochim. Acta* **2019**, *305*, 125–136. [CrossRef]
11. Yang, C.; Hu, L.; Zang, L.M.; Liu, Q.F.; Qiu, J.H.; Yang, J.; Qiao, X. High-Performance All-Solid-State Supercapacitor Based on Activated Carbon Coated Fiberglass Cloth Using Asphalt as Active Binder. *J. Electrochem. Soc.* **2020**, *167*, 20540. [CrossRef]

12. Liu, K.; Yao, Y.; Lv, T.; Li, H.L.; Li, N.; Chen, Z.L.; Qian, G.J.; Chen, T. Textile-like electrodes of seamless graphene/nanotubes for wearable and stretchable supercapacitors. *J. Power Sources* **2020**, *446*, 227355. [CrossRef]
13. Guo, Z.G.; Zhao, J.; Sun, C.; Cai, Z.H.; Ge, F.Y. Flexible self-standing carbon fabric electrode prepared by using simple route for wearable applications. *J. Mater. Sci. Mater. Electron.* **2020**, *31*, 1554–1565. [CrossRef]
14. Bharathidasan, P.; Subramaniam, T.; Chandini, D.; Sivakkumar, S.R.; Rajan, K.S.; Devaraj, S. Simultaneous Enhancement of Energy and Power Density of Reduced Graphene Oxide by the Effect of Dispersed Metal Oxide Nanoparticles in the Electrolyte. *J. Electrochem. Soc.* **2020**, *167*, 140524. [CrossRef]
15. Wu, P.; Cheng, S.; Yao, M.H.; Yang, L.F.; Zhu, Y.Y.; Liu, P.P.; Xing, O.; Zhou, J.; Wang, M.K.; Luo, H.W.; et al. A Low-Cost, Self-Standing $NiCo_2O_4$@CNT/CNT Multilayer Electrode for Flexible Asymmetric Solid-State Supercapacitors. *Adv. Funct. Mater.* **2017**, *27*, 1702160. [CrossRef]
16. Li, X.L.; Yuan, L.B.; Liu, R.; He, H.N.; Hao, J.N.; Lu, Y.; Wang, Y.M.; Liang, G.M.; Yuan, G.H.; Guo, Z.P. Engineering Textile Electrode and Bacterial Cellulose Nanofiber Reinforced Hydrogel Electrolyte to Enable High-Performance Flexible All-Solid-State Supercapacitors. *Adv. Energy Mater.* **2021**, *11*, 2003010. [CrossRef]
17. Yong, S.; Owen, J.; Beeby, S. Solid-State Supercapacitor Fabricated in a Single Woven Textile Layer for E-Textiles Applications. *Adv. Eng. Mater.* **2018**, *20*, 1700860. [CrossRef]
18. Yong, S.; Hillier, N.; Beeby, S. The influence of textile substrate on the performance of multilayer fabric supercapacitors. *J. Ind. Text.* **2021**, *50*, 1397–1408. [CrossRef]
19. Li, L.Y.; Lu, F.X.; Wang, C.; Zhang, F.L.; Liang, W.H.; Kuga, S.; Dong, Z.C.; Zhao, Y.; Huang, Y.; Wu, M. Flexible double-cross-linked cellulose-based hydrogel and aerogel membrane for supercapacitor separator. *J. Mater. Chem. A* **2018**, *6*, 24468–24478. [CrossRef]
20. Liu, M.T.; Turcheniuk, K.; Fu, W.B.; Yang, Y.; Liu, M.; Yushin, G. Scalable, safe, high-rate supercapacitor separators based on the Al_2O_3 nanowire Polyvinyl butyral nonwoven membranes. *Nano Energy* **2020**, *71*, 104627. [CrossRef]
21. Liu, Z.Y.; Ma, S.; Li, X.; Yang, H.; Xu, Z.L. Porous carbonaceous composite derived from $Mg(OH)_2$ pre-filled PAN based membrane for supercapacitor and dye adsorption application. *J. Solid State Chem.* **2019**, *277*, 493–501. [CrossRef]
22. Arthi, R.; Jaikumar, V.; Muralidharan, P. Development of electrospun PVdF polymer membrane as separator for supercapacitor applications. *Energy Sources Part A* **2022**, *44*, 2294–2308. [CrossRef]
23. Torvi, A.I.; Munavalli, B.B.; Naik, S.R.; Kariduraganavar, M.Y. Scalable fabrication of a flexible interdigital micro-supercapacitor device by in-situ polymerization of pyrrole into hybrid PVA-TEOS membrane. *Electrochim. Acta* **2018**, *282*, 469–479. [CrossRef]
24. Na, R.Q.; Zhang, X.R.; Huo, P.F.; Du, Y.L.; Huo, G.Z.; Zhu, K.; Wang, G.B. High performance disulfonated poly(arylene ether sulfone)/poly(ethylene oxide) composite membrane used as a novel separator for supercapacitor with neutral electrolyte and activated carbon electrodes. *High Perform. Polym.* **2017**, *29*, 984–993. [CrossRef]
25. Li, Z.M.; Wei, J.G.; Shan, F.; Yang, J.; Wang, X.L. PVDF/PMMA brushes membrane for lithium-ion rechargeable batteries prepared via preirradiation grafting technique. *J. Polym. Sci. Polym. Phys.* **2008**, *46*, 751–758. [CrossRef]
26. Ahmed, F.E.; Lalia, B.S.; Hashaikeh, R. A review on electrospinning for membrane fabrication: Challenges and applications. *Desalination* **2015**, *356*, 15–30. [CrossRef]
27. Holda, A.K.; Vankelecom, I.F.J. Understanding and guiding the phase inversion process for synthesis of solvent resistant nanofiltration membranes. *J. Appl. Polym. Sci.* **2015**, *132*, 42130. [CrossRef]
28. Meng, X.L.; Ji, Y.; Yu, G.H.; Zhai, Y.J. Preparation and Properties of Polyvinylidene Fluoride Nanocomposited Membranes based on Poly(N-Isopropylacrylamide) Modified Graphene Oxide Nanosheets. *Polymers* **2019**, *11*, 473. [CrossRef]
29. Jang, J.; Oh, J.; Jeong, H.; Kang, W.; Jo, C. A Review of Functional Separators for Lithium Metal Battery Applications. *Materials* **2020**, *13*, 4625. [CrossRef]
30. Peng, Y.; Yao, R.; Yang, W.S. A poly(amidoamine) nanoparticle cross-linked two-dimensional metal-organic framework nanosheet membrane for water purification. *Chem. Commun.* **2019**, *55*, 3935–3938. [CrossRef]
31. Wang, B.H.; Thukral, A.; Xie, Z.Q.; Liu, L.M.; Zhang, X.N.; Huang, W.; Yu, X.G.; Yu, C.J.; Marks, T.J.; Facchetti, A. Flexible and stretchable metal oxide nanofiber networks for multimodal and monolithically integrated wearable electronics. *Nat. Commun.* **2020**, *11*, 2405. [CrossRef]
32. Tomic, N.Z.; Marinkovic, A.D.; Veljovic, D.; Trifkovic, K.; Levic, S.; Radojevic, V.; Heinemann, R.J. A new approach to compatibilization study of EVA/PMMA polymer blend used as an optical fibers adhesive: Mechanical, morphological and thermal properties. *Int. J. Adhes. Adhes.* **2018**, *81*, 11–20. [CrossRef]
33. Cheng, S.K.; Chen, C.Y. Mechanical properties and strain-rate effect of EVA/PMMA in situ polymerization blends. *Eur. Polym. J.* **2004**, *40*, 1239–1248. [CrossRef]
34. Hillier, N.; Yong, S.; Beeby, S. Optimization of Carbon Electrodes for Solid-State E-Textile Supercapacitors. *J. Phys. Conf. Ser.* **2019**, *1407*, 12059. [CrossRef]
35. Hillier, N.; Yong, S.; Cruden, A.; Beeby, S. Acetonitrile-Free Organic Electrolyte for Textile Supercapacitor Applications. *J. Electrochem. Soc.* **2021**, *168*, 80520. [CrossRef]
36. Xu, D.M.; Teng, G.H.; Heng, Y.Q.; Chen, Z.Z.; Hu, D.Y. Eco-friendly and thermally stable cellulose film prepared by phase inversion as supercapacitor separator. *Mater. Chem. Phys.* **2020**, *249*, 122979. [CrossRef]

polymers

MDPI

Article

Modeling and Simulation of Mechanical Performance in Textile Structural Concrete Composites Reinforced with Basalt Fibers

Rajesh Kumar Mishra [1,*], Bijoya Kumar Behera [2], Vijay Chandan [1], Shabnam Nazari [1] and Miroslav Muller [1]

[1] Department of Material Science and Manufacturing Technology, Faculty of Engineering, Czech University of Life Sciences Prague, 16500 Prague, Czech Republic
[2] Department of Textile Engineering, Indian Institute of Technology, New Delhi 110016, India
* Correspondence: mishrar@tf.czu.cz

Abstract: This investigation deals with the prediction of mechanical behavior in basalt-fiber-reinforced concrete using the finite element method (FEM). The use of fibers as reinforcement in concrete is a relatively new concept which results in several advantages over steel-reinforced concrete with respect to mechanical performance. Glass and polypropylene (PP) fibers have been extensively used for reinforcing concrete for decades, but basalt fibers have gained popularity in recent years due to their superior mechanical properties and compatibility with concrete. In this study, the mechanical properties of basalt-fiber-reinforced concrete are predicted using FEM analysis, and the model results are validated by conducting experiments. The effect of fiber-volume fraction on the selected mechanical performance of concrete is evaluated in detail. Significant improvement is observed when the loading is increased. There are superior mechanical properties, e.g., load bearing and strain energy in basalt-fiber-reinforced concrete as compared to conventional concrete slabs reinforced with gravel or stones. The results of the simulations are correlated with experimental samples and show a very high similarity. Basalt-fiber-reinforced concrete (BFRC) offers a lightweight construction material as compared to steel-fiber-reinforced concrete (SFRC). Further, the problem of corrosion is overcome by using this novel fiber material in concrete composites.

Keywords: fiber-reinforced concrete; basalt; finite element analysis (FEA); stress; strain energy; ductility

1. Introduction

Fiber-reinforced composites have been utilized for quite some time now. Composites using fibrous reinforcement materials with high strength, low density, along with high ductility are used in many fields such as aerospace, construction, and defense. Due to their evident advantages, much research has been carried out in this field to examine the possibility of using such composites in other fields where they have not been used. This approach also helps in decreasing the cost of materials for more widespread applications. Fiber-reinforced concrete, as opposed to conventional concrete, is a relatively new concept. The use of fibers in concrete decreases the weight of the slabs and columns. They also have very good insulation properties and are not magnetic in nature. Though at lower loads the crack widths and extensions are higher for fiber-reinforced concrete due to its lower Young's modulus, at higher loads, crack spacing decreases relative to steel reinforcement. Creep and shrinkage behavior in fiber-reinforced concrete is similar to those in steel-reinforced concrete slabs and columns [1,2]. The fibers used for this purpose are glass, polypropylene (PP), basalt, and carbon fibers (though rarely, due to their relatively higher cost) [3–7].

Basaltic rock is formed from solidified lava (igneous rock). It is an extremely hard, fine-grained extrusive rock and consists of the minerals SiO_2, Al_2O_3, CaO, MgO, K_2O, Na_2O, Fe_2O_3 and FeO. It has high abrasion resistance, excellent resistance to extreme temperatures, good insulation properties (heat, sound, and electrical), lower moisture absorption, inertness to the environment, and very high durability [8–11]. Basalt fibers are extracted from this rock through melt extrusion. The rock is first broken into very small pieces, melted, and then made to flow through bushings with many fine orifices to form the fiber. This process does not use any additional materials; the fiber is made with a naturally available material and requires less energy and is cheaper than many other high-performance fibers such as glass, carbon, etc. The chemical composition of basalt fiber is very similar to that of glass fiber [12].

Unlike many other fibers, basalt fiber is known to properly disseminate in the concrete mix when used instead of forming balls and has very high bio-solubility. Its stress–strain behavior is linear up to its failure [13–17]. It is more economic compared with carbon fiber and is more advantageous compared with glass fiber with respect to many criteria (e.g., Young's modulus, acid resistance, etc.) and has therefore been extensively investigated in many studies in order to explore the possibility of replacing glass fibers in concrete [18–22].

Basalt-fiber-reinforced concrete shows higher splitting tensile strength and toughness as compared to pure concrete, but its compressive strength decreases over time. With increasing fiber length, there is an increase in tensile strength, compressive strength, and toughness, which may be because interfacial bonding increases with increasing surface area [23–26]. Basalt fiber shows a higher Young's modulus than glass fiber, and the compressive strength and flexural behavior of basalt-reinforced concrete is found to be superior. However, basalt fibers show lower tensile strength as compared to glass fibers [27]. Basalt fibers or fibrous materials are usually coated with a binding agent in order to make them compatible with the concrete. It has been observed that basalt fibers are 'wetted out' by these binders well [28–31]. A higher number of admixtures must be added in order to increase the slump to the required level. This is due to the fact that the fibers have larger surface areas as compared to the aggregates in the mix, and they bond with a higher amount of cement, thereby increasing the viscosity of the mix [32,33].

Basalt fibers are extracted from volcanic rocks and are therefore a purely natural product. They do not involve artificial chemicals or reagents. They have been referred to as a "green material" in the literature. The environmental impact of using such material instead of synthetic material is very significant [33,34].

Using chopped basalt fibers in concrete mix is not the only way of reinforcing. Basalt-woven structures or rebars are other alternatives. They are lighter than other forms of reinforcement, and using a lower amount of basalt fibers/rebars may generate the same strength while considerably reducing the weight of the slabs [34,35]. They can also be cut into the length required more easily with the normal available cutting tools.

In this study, the mechanical properties of basalt-fiber-reinforced concrete slabs were predicted using the finite element analysis. Simulations were conducted using ANSYS by making CAD models of the samples and the obtained results were analyzed. To study the tensile property of basalt-fiber-reinforced concrete, experiments were carried out by making samples of pure concrete and 1%, 2%, and 3% basalt-fiber-(by weight) reinforced concrete. The simulation results were compared with experimental samples as well as with findings of previous studies. The properties were also compared to those of fine and coarse aggregates used in concrete. The influence of basalt fiber as a replacement for stone/aggregates in concrete slabs was investigated.

2. Materials and Methods

2.1. Materials

Details of the basalt fibers and concrete used are given in Table 1. The basalt fiber was received from the company Basaltex (Wevelgem, Belgium). The standard commercial concrete was purchased from a local market.

Table 1. Specifications of the basalt fiber and concrete used.

Properties	Basalt Fiber	Concrete
Length	1 cm	
Diameter	0.5 mm	
Grade		M30
Density	2650 kg/m^3	2240 kg/m^3
Young's modulus	89 GPa	17 GPa
Poisson's ratio	0.3	0.33
Tensile yield strength	4560 MPa	0.142 MPa
Compressive yield strength	49.5 MPa	24.13 MPa
Ultimate tensile strength	4840 MPa	2.2 MPa
Ultimate compressive strength	45.326 MPa	40 MPa

The chemical composition of the basalt fiber is given in Table 2.

Table 2. Chemical composition of the basalt fiber used.

Chemical Composition of Basalt Rocks	%
SiO_2	52.8
Al_2O_3	17.5
Fe_2O_3	10.3
MgO	4.63
CaO	8.59
Na_2O	3.34
K_2O	1.46
TiO_2	1.38

A qualitative elemental analysis of the basalt fiber determined that SiO_2 and Al_2O were the dominant compounds. The content of FeO and Fe_2O_3 played a very important role in determining many physico-mechanical properties of basalt fibers, such as density, color (from brown to dull green, depending on the FeO content), lower heat conduction, and better temperature stability compared with those of E-glass fibers. The basalt fibers were pretreated against possible alkali reactions [33,35].

2.2. Methods

Experimental Samples

Four types of experimental samples were prepared as shown in Figure 1. They were the plain concrete sample, 1% (by weight of concrete) basalt-fiber-reinforced sample, 2% basalt-fiber-reinforced sample, and 3% basalt-fiber-reinforced sample. The concrete samples were prepared by mixing the cement-aggregate mixture with 30% water (by weight of concrete) and different percentages of chopped, short staple basalt fibers (by weight of concrete). The samples had a dimension of 11 cm $\times$ 7 cm $\times$ 1 cm. These samples were tested under a tensile load.

The tensile testing was conducted on a Zwick/Roell universal tensile tester as per the standard ASTM C1399 [36,37]. Twenty samples sized 11 cm $\times$ 7 cm $\times$ 1 cm were tested for each experiment. The average of 20 measurements was reported to have significance at a 95% confidence interval. The coefficient of variation was below 5%. The testing device is shown in Figure 2.

Figure 1. Samples prepared for tensile testing.

Figure 2. Zwick/Roell universal tensile tester.

The average material property was used for defining the composite/concrete elements. Multi-scale mechanical models were developed in order to predict the properties of the fiber-based composites.

For the theoretical estimation of the mechanical properties of the fiber-reinforced concrete, the corresponding mechanical properties of the fibers and pure concrete were experimentally evaluated. The mechanical properties of the basalt-fiber-reinforced concrete was estimated using a Halpin–Tsai model. This model is commonly used for the prediction of the effective mechanical properties of fiber-reinforced concrete (composites) [36–38]. The Halpin–Tsai equation is expressed as:

$$K_c = K_m \left[\frac{1 + \xi \zeta V_f}{1 - \eta V_f} \right] \tag{1}$$

$$\text{With } \eta = \left[\frac{\left(K_f/K_m\right) - 1}{\left(K_f/K_m\right) + \zeta} \right] \tag{2}$$

where,
K_c = the effective mechanical property of the fiber-reinforced concrete;
K_f and K_m are the corresponding fiber and concrete mechanical properties, respectively;
V_f = the fiber volume fraction;
ζ = the geometrical parameter, which represents the loading conditions (e.g., uniaxial, biaxial, multiaxial, etc.).

The modeling was carried out using a micro-scale representative volume element (RVE), which was subsequently transferred to a meso-scale RVE of the final composite to calculate the elastic constants by homogenization. Research has been reported about the prediction of mechanical properties in concrete with a given fiber-volume percentage. The predictions were reported to have a very good agreement with the experimental results [39,40].

The FE solution is broken down into the following three stages. This is the basic guideline that can be used for setting up any FEA [39–41].

The major steps in preprocessing are given below:

- Defining key points/lines/areas/volumes in the concrete slab.
- Defining element type and material/geometric properties for fiber, gravels, as well as the cementitious matrix.
- Defining mesh lines/areas/volumes as required.

The amount of detail required depends on the dimensionality of the analysis (i.e., 1D, 2D, axisymmetric, and 3D).

Simulations were conducted on the following models to test their tensile strength:

- Plain concrete sample;
- Concrete sample reinforced with 1% basalt fiber;
- Concrete sample reinforced with 2% basalt fiber;
- Concrete sample reinforced with 3% basalt fiber.

In addition to these, some other simulations were also conducted in order to compare the fiber-reinforced concrete with the standard concrete reinforced with aggregates/stones. For such a simulation, the volume fraction of aggregates was considered as 3%. The density of such aggregates was taken from the literature [31,34,38]. The properties of concrete and the aggregates were inputted into the model. All other conditions of loading were the same for the 3% basalt-fiber-reinforced concrete. The parameters were selected as follows:

- Three percent very fine aggregates (average 4 mm) with a density of 2615 kg/m^3 and a volume equivalent to that of the chopped fibers;
- Three percent coarse aggregates (average 7 mm) with a density of 2660 kg/m^3 and a volume equivalent to that of the chopped fibers.

A unit cell was prepared for each case with a dimension of $0.5 \times 7 \times 1$ cm. A total of twenty-two (22) unit cells were stacked one above the other to make the complete concrete slab structures shown in Figures 3–8.

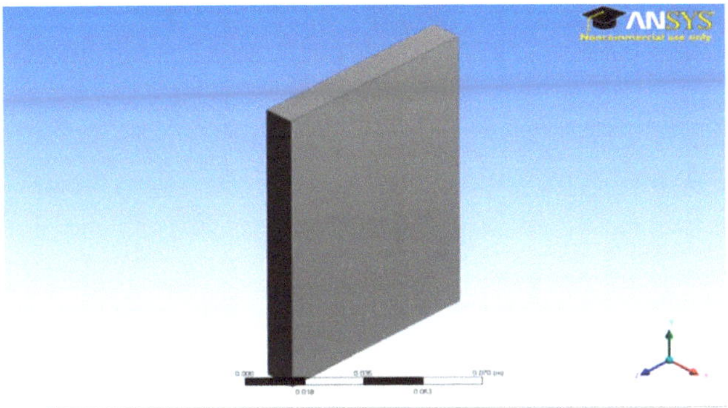

Figure 3. Plain concrete slab model (11 × 7 × 1 cm).

Figure 4. One percent fiber sample model (11 × 7 × 1 cm) and concrete slab.

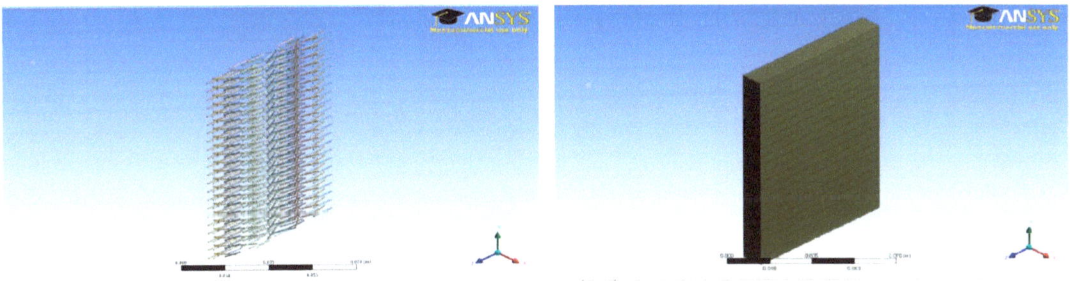

Figure 5. Two percent fiber sample model (11 × 7 × 1 cm) and concrete slab.

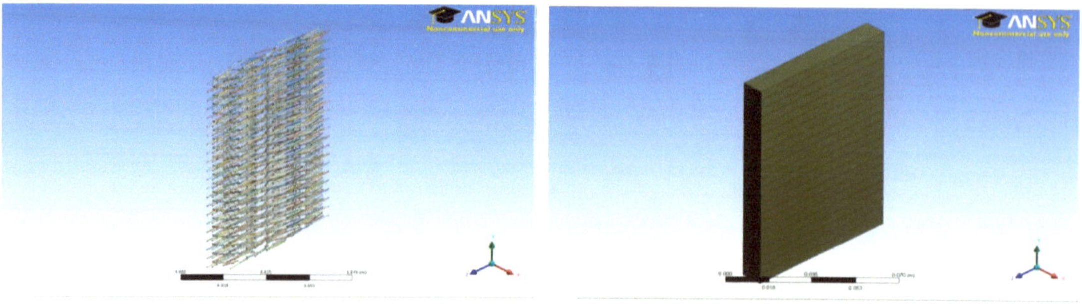

Figure 6. Three percent fiber sample model (11 × 7 × 1 cm) and concrete slab.

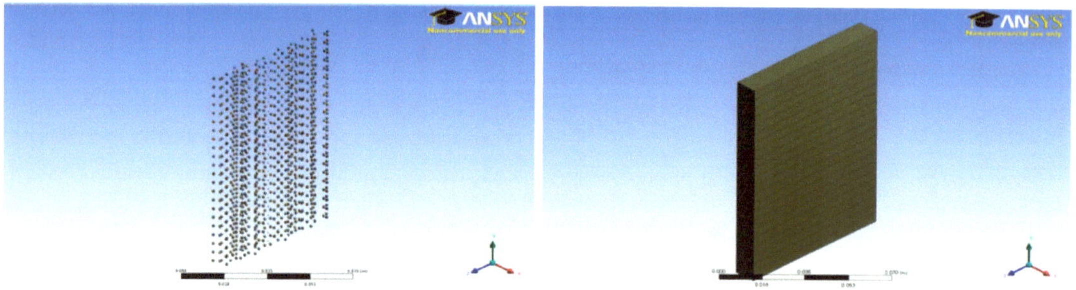

Figure 7. Three percent fine aggregate sample model (11 × 7 × 1 cm) and concrete slab.

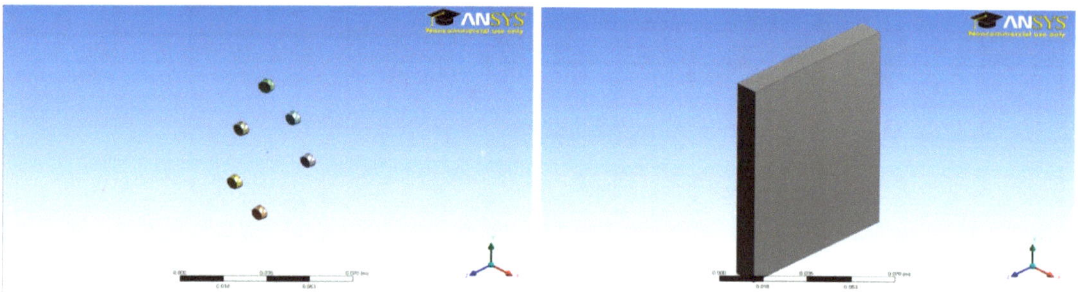

Figure 8. Three percent coarse aggregate sample model (11 × 7 × 1 cm) and concrete slab.

The static structural module in ANSYS was used to carry out the simulation in order to compare the tensile strengths of each of the samples. The following simulations were carried out:

- The simulation of all samples with one end fixed and a force of 50×10^5 N acting on the opposite face.
- The simulation of the pure concrete sample with one end fixed and a force of 150×10^5 N acting on the opposite face (the force at which pure concrete the block reached failure in the actual conducted experiment).
- The simulation of the 1% fiber-reinforced concrete sample with one end fixed and a force of 300×10^5 N acting on the opposite face (the force at which the 1% fiber-reinforced concrete sample failed in the conducted experiment).

- The simulation of the 2% fiber sample with one end fixed and a force of 500×10^5 N acting on the opposite face (the force at which the 2% fiber-reinforced concrete sample failed in the conducted experiment).
- The simulation of all the 3% filler-based concrete samples (fibers, fine aggregates, and coarse aggregates) with one end fixed and a force of 700×10^5 N acting on the opposite face (the force at which the 3% fiber sample failed in the experiment conducted).

To determine the advantages of the linear behavior of basalt fiber until failure, simulations were also conducted assuming a yield point before the ultimate failure. The simulations for these are not shown in this paper, but the effects were insignificant.

The input parameters given are as follows:

Young's modulus of basalt fiber = 89 GPa; Poisson's ratio of basalt fiber = 0.3.

Young's modulus of concrete = 17 GPa; Poisson's ratio of concrete = 0.33.

The tensile yield and ultimate strength values, as well as the compression yield and ultimate strengths were also given as inputs, but they were found to have no effect on the results as explained later in this section.

The basalt fibers, fine aggregates, and coarse aggregates were all given with the same dimensions for comparison purposes. The model was built with a coarse mesh and the simulation was carried out. The distribution of stress along the samples were found for the following parameters in each sample:

Directional deformation (m)—(along the y-axis, along which the force was applied).

Strain energy (U) is given by the following expression:

$$U = 0.5 \times (V/E) \times S^2 \tag{3}$$

where U = strain energy, V = volume, E = Young's modulus, and S = stress.

This gave a measurement of Young's modulus of different samples under similar stress conditions. The higher the stored strain energy, the lower the Young's modulus of the structure (or part of the structure being analyzed).

3. Results and Discussion

3.1. Experimental Results

The experimental samples were tested on a universal testing machine. The average result of 20 measurements was taken. The load extension graph obtained for the samples is given in Figure 9.

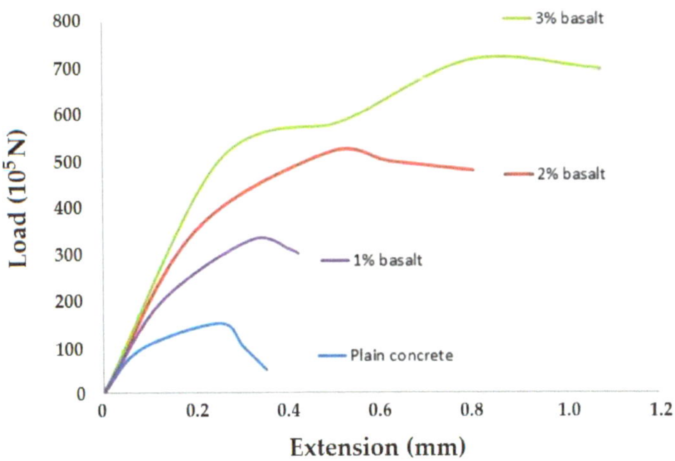

Figure 9. Load-extension curves from tensile testing of experimental concrete slabs.

The tensile strength significantly increased as the percentage of the fiber increased. The strength increased by around four times when the fiber percentage was 2%. The elastic modulus also increased when the fiber percentage increased from 0 to 1% but after that, remained almost constant when further increased. The crack patterns were also very different for the sample without the fibers and the samples reinforced with the fibers. The failure/crack path in the case of the samples reinforced with basalt fibers was much more tortuous due to the presence of fibers, not a straight horizontal line (i.e., the shortest path), unlike the case of the pure concrete sample. The overall extension increased for the fiber-reinforced concrete, and it was also proportional to the fiber weight percentage. The samples are shown in Figure 10. The cracking pattern showed that with an increasing weight percentage of basalt fiber, cracking was more difficult. This indicated that basalt fibers are able to more efficiently absorb the tensile stress in concrete.

Figure 10. Crack patterns in 0%, 1%, 2%, and 3% basalt-fiber-reinforced concrete samples after tensile testing (pictures formatted so that crack patterns are clearer).

The peak load and extension are also given in Table 3.

Table 3. Experimental results for the mechanical behavior of concrete.

Type of Concrete Slab	Experimental Stress (10^5 Pa)	Deformation at Peak Stress (mm)
Pure concrete slab	0.52	0.75
1% basalt-fiber-reinforced concrete slab	3.64	0.67
2% basalt-fiber-reinforced concrete slab	6.11	0.22
3% basalt-fiber-reinforced concrete slab	7.68	0.13
3% fine-aggregate-reinforced concrete slab	2.65	0.31
3% coarse-aggregate-reinforced concrete slab	5.52	0.51

It can be observed that an addition of basalt fiber of up to 3% increased the peak stress in the concrete. Further, the deformation at peak stress was reduced, thus increasing the load-bearing capacity. The basalt-fiber-reinforced concrete showed improved performance as compared to pure concrete as well as fine-/coarse-aggregate-based samples. Similar studies have been reported in the literature where the performance of concrete was improved by an addition of up to 3% microfibers and waste mineral admixtures [42].

The inclusion of micro-silica in small doses has been reported to be superior with respect to the tensile performance of concrete. It has also been reported that micro-fillers enhance the bond strength, which was responsible for the improvement of mechanical performance in concrete composites [43].

3.2. Results of Simulations Using FEM

The simulations of concrete samples were carried out based on the defined conditions. The results are shown in Figures 11–22.

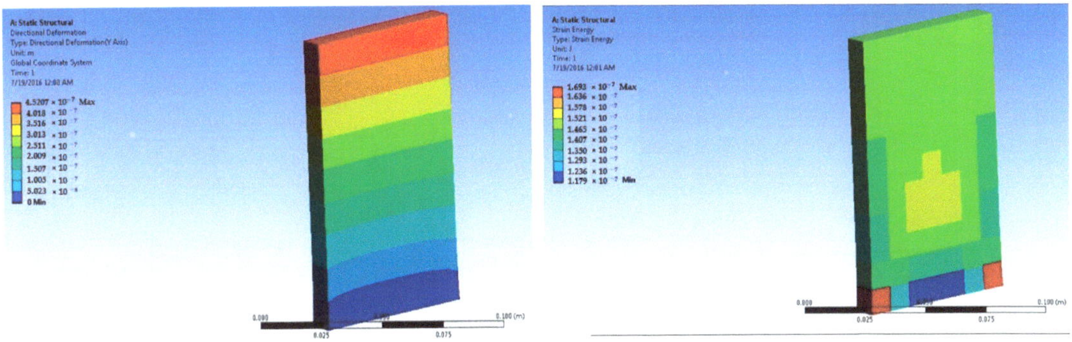

Figure 11. Plain concrete slab at 50 × 10^5 N force (deformation and strain energy).

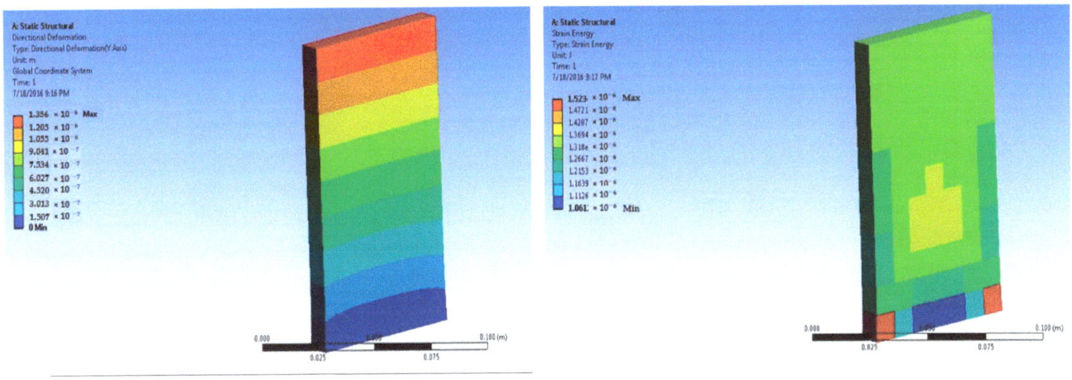

Figure 12. Plain concrete slab at 150 × 10^5 N force (deformation and strain energy).

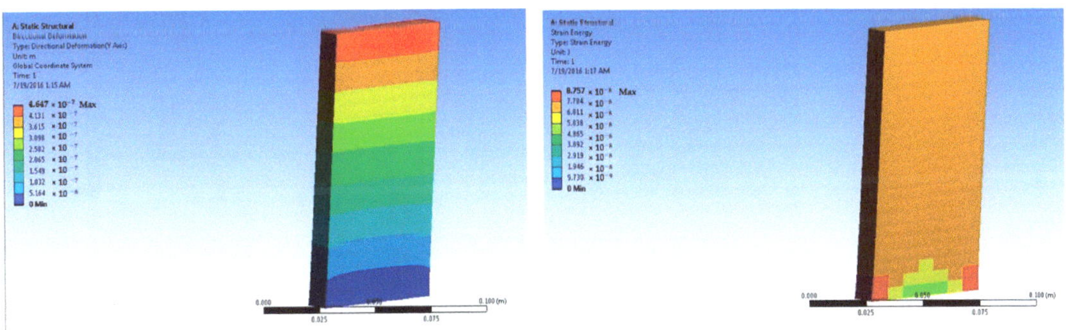

Figure 13. One percent basalt-fiber-reinforced concrete slab at 50 × 10^5 N force (deformation and strain energy).

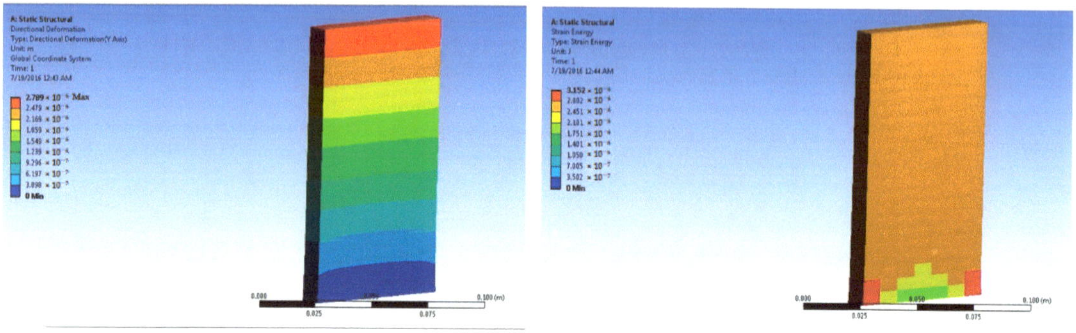

Figure 14. One percent basalt-fiber-reinforced concrete slab at 300 × 10^5 N force (deformation and strain energy).

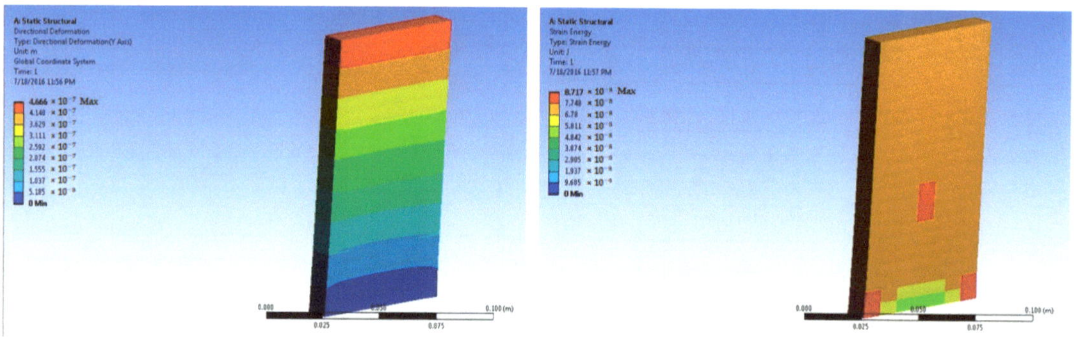

Figure 15. Two percent basalt-fiber-reinforced concrete slab at 50 × 10^5 N force (deformation and strain energy).

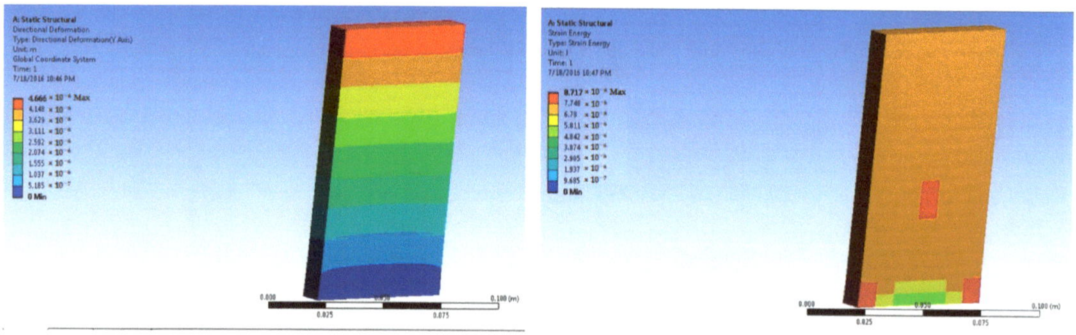

Figure 16. Two percent basalt-fiber-reinforced concrete slab at 500 × 10^5 N force (deformation and strain energy).

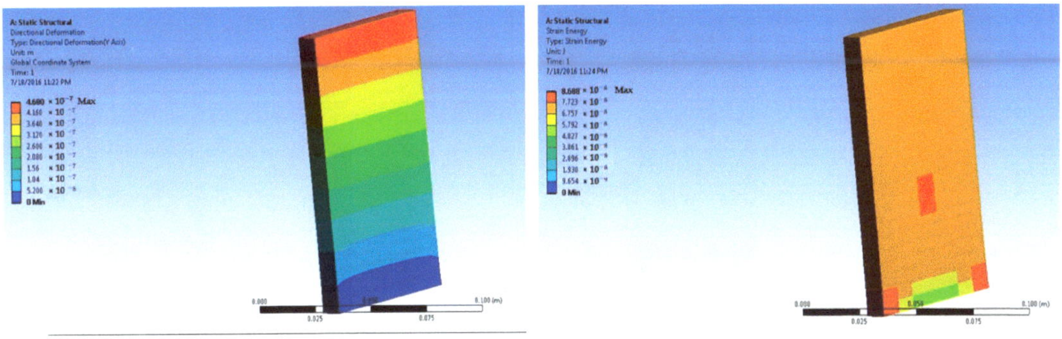

Figure 17. Three percent basalt-fiber-reinforced concrete slab at 50 × 10^5 N force (deformation and strain energy).

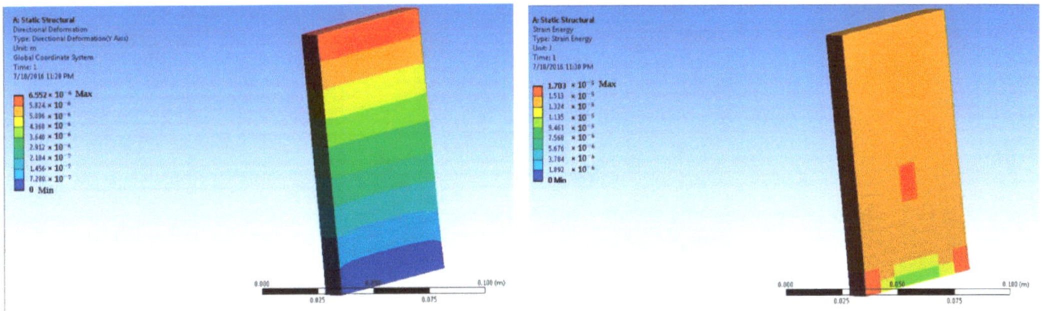

Figure 18. Three percent basalt-fiber-reinforced concrete slab at 700 × 10^5 N force (deformation and strain energy).

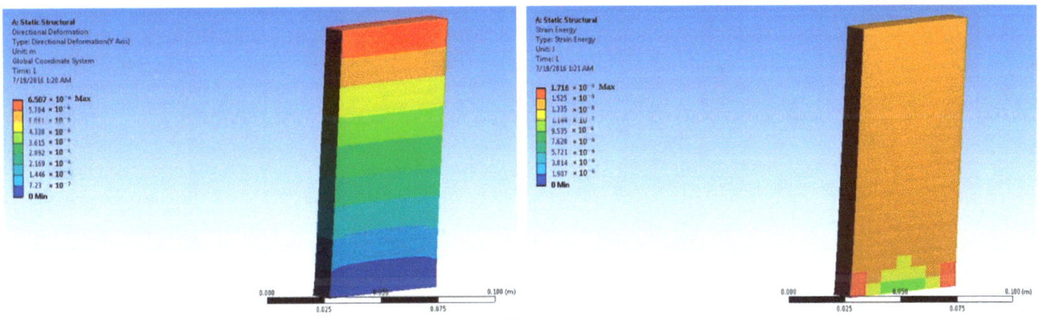

Figure 19. Three percent fine-aggregate-based concrete slab at 50 × 10^5 N force (deformation and strain energy).

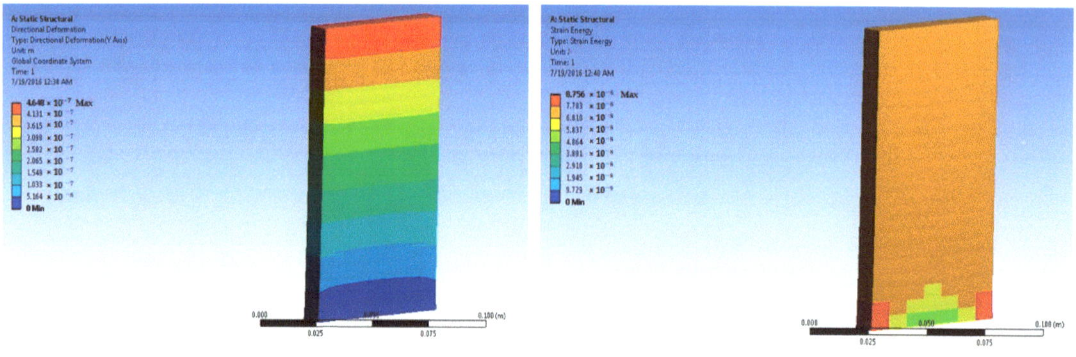

Figure 20. Three percent fine-aggregate-based concrete slab at 700 × 10^5 N force (deformation and strain energy).

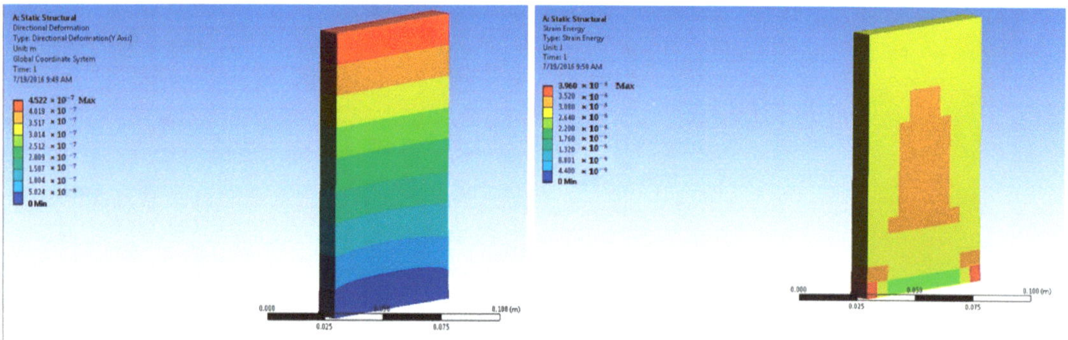

Figure 21. Three percent coarse-aggregate-based concrete slab at 50 × 10^5 N force (deformation and strain energy).

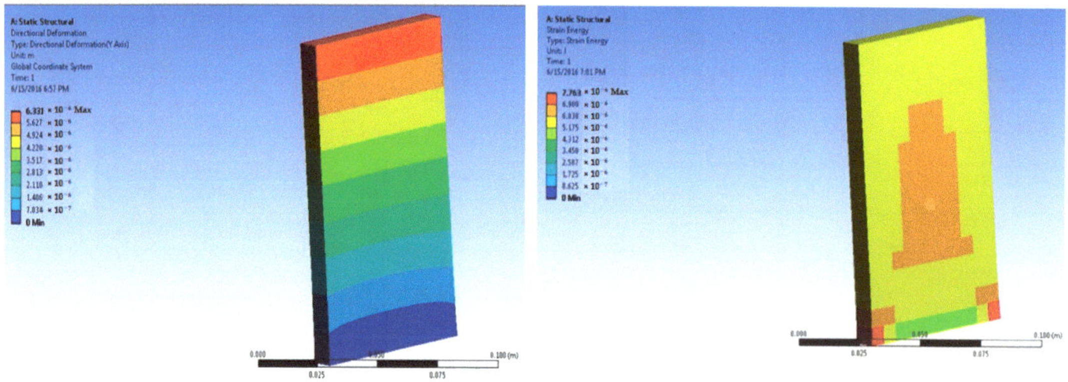

Figure 22. Three percent coarse-aggregate-based concrete slab at 700 × 10^5 N force (deformation and strain energy).

The model for the plain concrete sample predicted a maximum stress of 6.9×10^4 Pa and a maximum strain 4.2×10^{-6} m/m at 50×10^5 N force. The corresponding deformation was 1.36×10^{-6} m. These values were quite close to the experimental result.

The predicted value for maximum stress at 150×10^5 N force was 2.1×10^5 Pa. The maximum strain was predicted as 1.25×10^{-5} m/m.

The model for the 1% basalt-fiber-reinforced concrete predicted an increased maximum stress at 50×10^5 N force. The value was predicted to be 4.35×10^5 Pa. The strain was predicted as 4.75×10^{-6} m/m.

Simulations were performed for 1% basalt-fiber-reinforced concrete at a load of 300×10^5 N. they were based on the peak load obtained during the experimental trials. The predicted value of maximum stress was 4.35×10^5 Pa. The corresponding strain was predicted as 2.9×10^{-5} m/m.

Further simulations were performed for the 2% basalt-fiber-reinforced concrete samples at 50×10^5 N force. The predicted maximum stress was 7.29×10^4 Pa. The maximum strain was predicted as 4.76×10^{-6} m/m. The performance was predicted as superior to that of the 1% basalt-fiber-reinforced concrete sample. This also corresponded to the findings of the experimental analysis.

Based on the peak experimental load, the simulations for the 2% basalt-fiber-based concrete were performed at 500×105 N force. The maximum stress was predicted as 7.22×10^5 Pa. This was significantly higher than the maximum stress for the 1% fiber-loaded concrete. Further, the stress was higher at a higher load. The maximum strain was predicted as 4.76×10^{-5} m/m. It was observed to be slightly higher at higher loading. The performance was similar to that of the results obtained from the experimental samples.

Simulations for the 3% fiber-reinforced sample were carried out at 50×10^5 N force. The maximum stress was predicted as 7.38×10^4 Pa. This was much higher than the stress level for the 1% and 2% fiber-loaded samples. These findings resembled the experimental tests.

Based on the peak experimental load, further simulations were performed for the 3% basalt-fiber-reinforced concrete slabs at a load of 700×10^5 N. The maximum stress was predicted as 10.18×10^5 Pa. This value was significantly higher than the maximum stress for the 1% and 2% basalt-reinforced concrete. The corresponding strain was predicted as 6.66×10^{-5} m/m.

For a comparison of the fiber-reinforced concrete with respect to the standard concrete with fine aggregates, simulations were performed at 50×10^5 N force. The maximum stress was predicted as 7.05×10^4 Pa. This stress value was higher than that of the plain concrete but lower than that of even the 1% basalt-fiber-based concrete. Further, the strain was predicted as 4.42×10^{-6} m/m, which was lower than that of the basalt-fiber-based samples but slightly higher than that of the plain concrete sample.

The fine-aggregate-based sample was also loaded at 700×10^5 N force in order to compare its performance with the corresponding sample of the 3% basalt-fiber-reinforced concrete slab. It was predicted that the maximum stress would be 7.12×10^5 Pa. This was higher than the maximum stress for the 1% basalt-fiber-based sample. However, it was significantly lower than that of the 3% basalt-fiber-reinforced sample and slightly lower than that of the 2% fiber-based sample. This was an indication that a 2–3% reinforcement of basalt fiber in the concrete would be optimum. It was also established through the experimental analysis that the 3% basalt-fiber-based samples showed a higher stress level than that of the 3% fine-aggregate-based concrete [44,45]. A higher fiber content resulted in several practical and workability problems. The corresponding strain was predicted as 4.26×10^{-5} m/m. This was lower than the predicted stress for the basalt-fiber-reinforced concrete. This was the result of fiber extensibility, which was higher than that of the stones/aggregates.

Simulations were performed on concrete samples reinforced with coarse aggregates. The loading condition was set as 50×10^5 N. The maximum stress was predicted as 6.96×10^4 Pa, which was lower than that of all the samples reinforced with basalt fiber.

The maximum strain was predicted as 4.31×10^{-6} m/m. Thus, coarse aggregates could be successfully replaced with basalt fibers comprising up to 3% of the overall volume in a concrete mix.

The sample with 3% coarse aggregates was also simulated under 700×10^5 N force. The maximum stress of 6.58×10^5 Pa was predicted. It was significantly lower than the maximum stress in samples reinforced with 2% and 3% basalt fibers. Although, it was higher than that of the pure concrete sample. This indicated that coarse aggregates can be useful for the reinforcement of concrete; however, much better performances can be achieved by replacing these with basalt fibers comprising up to 3% of the overall volume. The maximum strain for the coarse-aggregate-based sample was predicted as 3.88×10^{-5} m/m. This was smaller than that of the basalt-fiber-based samples with 2% and 3% loading. This was an indication of higher deformability under loading when there was 2% and 3% basalt fiber in concrete. Such behavior in basalt-fiber-reinforced concrete can prove beneficial under severe loading conditions, e.g., in the case of earthquakes. Similar results have been reported in the literature that a hybrid effect of fibers and fillers improved the crack resistance in reinforced composites and concretes. This performance was dependent on the dosing percentage of the micro-fillers [44,45].

The results of the simulation at 50×10^5 N tensile force are summarized in Table 4.

Table 4. Simulation results for mechanical behavior of concrete at 50×10^5 N tensile force.

Type of Concrete Slab	Maximum Deformation (m)	Strain Energy (J)	Normal Stress (Pa)	Normal Strain (m/m)
Pure concrete slab	4.52×10^{-7}	1.53×10^{-7}	6.9×10^4	4.2×10^{-6}
1% basalt-fiber-reinforced concrete slab	4.64×10^{-7}	7.79×10^{-8}	7.2×10^4	4.75×10^{-6}
2% basalt-fiber-reinforced concrete slab	4.67×10^{-7}	7.75×10^{-8}	7.29×10^4	4.76×10^{-6}
3% basalt-fiber-reinforced concrete slab	4.68×10^{-7}	7.74×10^{-8}	7.38×10^4	4.76×10^{-6}
3% fine-aggregate-reinforced concrete slab	4.65×10^{-7}	8.76×10^{-8}	7.05×10^4	4.42×10^{-6}
3% coarse-aggregate-reinforced concrete slab	4.52×10^{-7}	3.96×10^{-8}	6.96×10^4	4.31×10^{-6}

The results of the simulation based on the corresponding load values obtained from the experimental tensile tests are summarized in Table 5.

Table 5. Simulation results of mechanical behavior of concrete at tensile failure load (load value obtained from the experiment value of peak load just before failure).

Type of Concrete Slab	Maximum Deformation (m)	Strain Energy (J)	Normal Stress (Pa)	Normal Strain (m/m)
Pure concrete slab (150×10^5 N)	1.36×10^{-6}	1.32×10^{-6}	2.1×10^5	1.25×10^{-5}
1% basalt-fiber-reinforced concrete slab (300×10^5 N)	2.79×10^{-6}	2.85×10^{-6}	4.35×10^5	2.9×10^{-5}
2% basalt-fiber-reinforced concrete slab (500×10^5 N)	4.67×10^{-6}	7.8×10^{-6}	7.22×10^5	4.76×10^{-5}
3% basalt-fiber-reinforced concrete slab (700×10^5 N)	6.55×10^{-6}	10.52×10^{-6}	10.18×10^5	6.66×10^{-5}
3% fine-aggregate-reinforced concrete slab (700×10^5 N)	6.51×10^{-6}	10.72×10^{-6}	7.12×10^5	4.26×10^{-5}
3% coarse-aggregate-reinforced concrete slab (700×10^5 N)	6.33×10^{-6}	7.76×10^{-6}	6.58×10^5	3.88×10^{-5}

The meshing used for the FEM was relatively coarse (because of the limitation of the number of nodes for which computation could be performed on the software). Although this did not have much effect on the amount of deformation, the other parameters (strain energy, normal stress, and normal strain) may have varied because of the difference in meshing, and thus the values may have been more accurate and precise. But it can be clearly observed that the trend was similar and independent of the meshing density/intensity. The conditions given in the experiment and the simulations were similar; therefore, the values obtained through the simulation were validated with the obtained experimental values. Thus, it is important to note that the trends were found to be similar to those shown in Figure 23. This indicated that both the computational simulation as well as the experiments were accurately conducted. The validation of the predicted results with the experimental analysis was successful. The error was limited to a maximum of 3%.

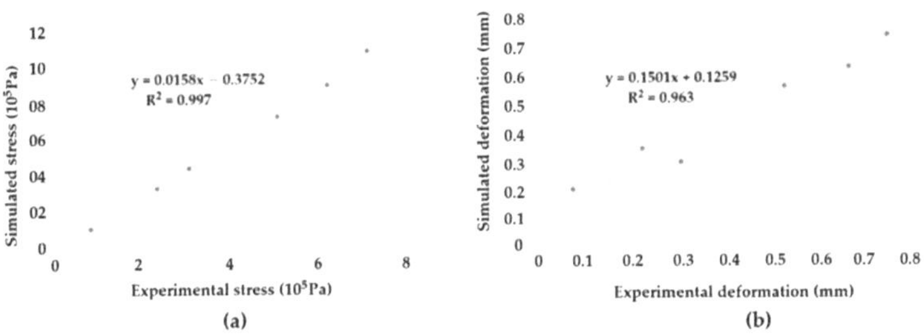

Figure 23. Correlation of results from simulation and experimental samples (**a**) stress and (**b**) deformation.

It can be observed from the results of the simulation that regarding 50×10^5 N force, the deformations in all the samples were almost the same (as in the case of the experimental samples). The Young's modulus increased between the 0% and 1% basalt-fiber-based samples but remained almost the same for the 2% and 3% fiber-reinforced samples. The same behavior was also observed from the experiments (the strain energies were compared to compare the Young's moduli). The force just-before-failure can also be noted (i.e., 150×10^5 N for pure concrete, 300 N for 1% fiber concrete, 500×10^5 N for 2% fiber, and 700×10^5 N for 3% fiber-based concrete). Maximum deformation was obtained for the 3% basalt-fiber-reinforced concrete, followed by the 2% fiber, 1% fiber, and pure concrete slabs. Therefore, it can be concluded that the values significantly varied when the fiber weight percentage was increased from 0% to 1%, but after that, they remained fairly similar, especially for those of the 2% and 3% basalt-fiber-reinforced samples (from 50×10^5 N force).

Other simulations were conducted in order to understand the influence of fillers, e.g., 3% basalt fiber, 3% fine aggregates, or 3% coarse aggregates. The behaviors of basalt-fiber- or fine-aggregate-reinforced concrete slabs were very similar. However, the 3% coarse-aggregate-reinforced sample showed very different results (they were to some extent similar to those of the pure concrete slabs in the 50×10^5 N force category). This seemed to prove that the basalt fiber distribution in the concrete mix influenced the strength, but the fiber structure itself did not affect the mechanical properties. The simulations for the 3% coarse-aggregate-reinforced concrete were conducted using a finer mesh, since the results were not obtained for the same coarse mesh used for other samples, but the values were comparable with those obtained for the pure concrete slabs with fine mesh. There was not much improvement of mechanical performance in the case of the coarse aggregates.

The tensile ultimate and yield points and compression ultimate and yield points of both fiber and cement were also given as input parameters, but it was observed that they did not have much influence on the obtained results.

4. Conclusions

The experimental results showed that increasing the percentage of basalt fiber in concrete increased the tensile strength by a significant amount and also improved the crack propagation patterns/tortuosity (it had better elastic behavior). From the simulations, it was also established that the percentage of basalt fiber (or aggregates) did not seem to make much difference when the percentage was increased from 1% to 3% (especially between 2% to 3%), but the increase from 0% to 1% significantly influenced the mechanical behavior. Maximum stress was predicted for 3% basalt-fiber-reinforced concrete samples. The results were found to be similar to those reported in the literature in this area. The simulations were further validated with experimental samples with relatively higher accuracy. The error of prediction was limited to 3%. Moreover, the distribution of the filler material (fiber or aggregates) in the sample influenced the mechanical properties, as observed from the 3% coarse aggregate and pure concrete samples. The fiber's internal structure itself did not have much effect on the mechanical properties of the concrete sample (shown by similar results for fine aggregates and fibers). The simulation software only considered the values of Young's modulus and Poisson's ratio for fiber-reinforced concrete. Changing the input parameters into ultimate tensile and compression strength or yield point strength did not affect the results.

Author Contributions: R.K.M., B.K.B., V.C., S.N. and M.M., experiment design; R.K.M., B.K.B., V.C., S.N. and M.M., testing of mechanical properties and data analysis; R.K.M., B.K.B., V.C., S.N. and M.M., writing and editing of the paper; R.K.M., B.K.B., V.C., S.N. and M.M., language correction; R.K.M., communication with the editors; R.K.M., B.K.B. and M.M., administration and funding; R.K.M., B.K.B. and M.M., supervision. All authors have read and agreed to the published version of the manuscript.

Funding: This work was supported by the Internal Grant Agency of the Faculty of Engineering, Czech University of Life Sciences, Prague, for the project "Numerical modeling and computational analyses of lightweight composite structures in automotive components" (no. 2022:31140/1312/3111).

Institutional Review Board Statement: Not applicable.

Informed Consent Statement: Not applicable.

Data Availability Statement: Not applicable.

Conflicts of Interest: The authors declare no conflict of interest.

References

1. Sun, X.; Gao, Z.; Cao, P.; Zhou, C. Mechanical properties tests and multiscale numerical simulations for basalt fiber reinforced concrete. *Constr. Build. Mater.* **2019**, *202*, 58–72. [CrossRef]
2. Mishra, R.; Tiwari, R.; Marsalkova, M.; Behera, B.K. Effect of TiO_2 nanoparticles on basalt/polysiloxane composites: Mechanical and thermal characterization. *J. Text. Inst.* **2012**, *103*, 1361–1368. [CrossRef]
3. Jamshaid, H.; Mishra, R.; Hussain, S. Basalt hybrid woven textile materials for advanced thermal applications. *Ind. J. Fibre Text. Res.* **2019**, *44*, 56–64.
4. Zhao, R.Y.; Wang, L.; Lei, Z.K.; Han, X.F.; Shi, J.N. Study on bending damage and failure of basalt fiber reinforced concrete under freeze-thaw cycles. *Constr. Build. Mater.* **2018**, *163*, 460–470. [CrossRef]
5. Mishra, R.; Jamshaid, H. Thermo-mechanical characteristics of basalt hybrid & non-hybrid woven fabric reinforced epoxy composites. *Polym. Compos.* **2016**, *37*, 2982–2994. [CrossRef]
6. Lago, B.D.; Taylor, S.E.; Deegan, P.; Ferrara, L.; Sonebi, M.; Crosset, P.; Pattarini, A. Full-scale testing and numerical analysis of a precast fibre reinforced self-compacting concrete slab pre-stressed with basalt fibre reinforced polymer bars. *Compos. Part B: Eng.* **2017**, *128*, 120–133. [CrossRef]
7. Jamshaid, H.; Mishra, R. Interfacial performance and durability of textile reinforced concrete. *J. Text. Inst.* **2018**, *109*, 879–890. [CrossRef]
8. Li, Y.; Zhang, J.; He, Y.; Huang, G.; Li, J.; Niu, Z.; Gao, B. A review on durability of basalt fiber reinforced concrete. *Compos. Sci. Technol.* **2022**, *225*, 109519. [CrossRef]
9. Haido, J.H. Flexural behavior of basalt fiber reinforced concrete beams: Finite element simulation with new constitutive relationships. *Structures* **2020**, *27*, 1876–1889. [CrossRef]
10. Behera, B.K.; Pattanayak, A.K.; Mishra, R. Prediction of fabric drape behavior using finite element method. *J. Text. Eng.* **2008**, *54*, 103–110. [CrossRef]
11. Mishra, R. Meso-scale finite element modeling of triaxial woven fabrics for composite in-plane reinforcement properties. *Text. Res. J.* **2013**, *83*, 1836–1845. [CrossRef]
12. Wang, Y.; Hu, S.; Sun, X. Experimental investigation on the elastic modulus and fracture properties of basalt fiber–reinforced fly ash geopolymer concrete. *Constr. Build. Mater.* **2022**, *338*, 127570. [CrossRef]
13. Xie, H.; Yang, L.; Zhang, Q.; Huang, C.; Chen, M.; Zhao, K. Research on energy dissipation and damage evolution of dynamic splitting failure of basalt fiber reinforced concrete. *Constr. Build. Mater.* **2022**, *330*, 127292. [CrossRef]
14. Mishra, R. Impact tolerance of 3D woven nanocomposites: A simulation approach. *J. Text. Inst.* **2013**, *104*, 562–570. [CrossRef]
15. An, H.; Song, Y.; Liu, L.; Meng, X. Experimental Study of the Compressive Strengths of Basalt Fiber-Reinforced Concrete after Various High-Temperature Treatments and Cooling in Open Air and Water. *Appl. Sci.* **2021**, *11*, 8729. [CrossRef]
16. Deng, Z.; Liu, X.; Liang, N.; de la Fuente, A.; Peng, H. Flexural Performance of a New Hybrid Basalt-Polypropylene Fiber-Reinforced Concrete Oriented to Concrete Pipelines. *Fibers* **2021**, *9*, 43. [CrossRef]

17. Mishra, R.; Gupta, N.; Pachauri, R.; Behera, B.K. Modeling and simulation of earthquake resistant 3D woven textile structural concrete composites. *Compos. Part B-Eng.* **2015**, *81*, 91–97. [CrossRef]
18. Mohammad Jani, N.; Shakir Nasif, M.; Shafiq, N.; Holt, I. Experimental Investigation on the Effect of Varying Fiber Mix Proportion on the Mechanical and Thermal Performances of Fiber-Reinforced Self-Compacting Concrete under Hydrocarbon Fire Condition. *Appl. Sci.* **2020**, *10*, 4586. [CrossRef]
19. Dvorkin, L.; Bordiuzhenko, O.; Tekle, B.H.; Ribakov, Y. A Method for the Design of Concrete with Combined Steel and Basalt Fiber. *Appl. Sci.* **2021**, *11*, 8850. [CrossRef]
20. Bakis, C.; Bank, L.C.; Brown, V.; Cosenza, E.; Davalos, J.F.; Lesko, J.J.; Triantafillou, T.C. Fiber-reinforced polymer composites for construction-state-of-the-art review. *J. Compos. Constr.* **2002**, *6*, 73–87. [CrossRef]
21. Jiang, C.; Fan, K.; Wu, F.; Chen, D. Experimental study on the mechanical properties and microstructure of chopped basalt fiber reinforced concrete. *Mater. Des.* **2014**, *58*, 187–193. [CrossRef]
22. Ayub, T.; Shafiq, N.; Nuruddin, M.F. Mechanical properties of high-performance concrete reinforced with basalt fibers. *Procedia Eng.* **2014**, *77*, 131–139. [CrossRef]
23. Ludovico, M.D.; Prota, A.; Manfredi, G. Structural upgrade using basalt fibers for concrete confinement. *J. Compos. Constr.* **2010**, *14*, 541–552. [CrossRef]
24. Lopresto, V.; Leone, C.; Iorio, I.D. Mechanical characterization of basalt fiber reinforced plastic. *Compos. Part B-Eng.* **2011**, *42*, 717–723. [CrossRef]
25. Kabay, N. Abrasion resistance and fracture energy of concretes with basalt fiber. *Constr. Build. Mater.* **2014**, *50*, 95–101. [CrossRef]
26. Manikandan, V.; Jappes, J.W.; Kumar, S.S.; Amuthakkannan, P. Investigation of the effect of surface modifications on the mechanical properties of basalt fiber reinforced polymer composites. *Compos. Part B-Eng.* **2012**, *43*, 812–818. [CrossRef]
27. Carmisciano, S.; Rosa, I.M.D.; Sarasini, F.; Tamburrano, A.; Valente, M. Basalt woven fiber reinforced vinylester composites: Flexural and electrical properties. *Mater. Des.* **2011**, *32*, 337–342. [CrossRef]
28. Sim, J.; Park, C. Characteristics of basalt fiber as a strengthening material for concrete structures. *Compos. Part B-Eng.* **2005**, *36*, 504–512. [CrossRef]
29. Urbanski, M.; Lapko, A.; Garbacz, A. Investigation on concrete beams reinforced with basalt rebars as an effective alternative of conventional R/C structures. *Procedia Eng.* **2013**, *57*, 1183–1191. [CrossRef]
30. Colombo, C.; Vergani, L.; Burman, M. Static and fatigue characterization of new basalt fiber reinforced composites. *Compos. Struct.* **2012**, *94*, 1165–1174. [CrossRef]
31. Wei, B.; Cao, H.; Song, S. Tensile behavior contrast of basalt and glass fibers after chemical treatment. *Mater. Des.* **2010**, *31*, 4244–4250. [CrossRef]
32. Wei, B.; Cao, H.; Song, S. Degradation of basalt fiber and glass fiber/epoxy resin composites in seawater. *Corros. Sci.* **2011**, *53*, 426–431. [CrossRef]
33. Mishra, R.; Jamshaid, H. A green material from rock: Basalt fiber- a Review. *J. Text. Inst.* **2016**, *107*, 923–937. [CrossRef]
34. Mishra, R.; Behera, B.K.; Banthia, V. Modeling and simulation of 3D orthogonal fabrics for composite applications. *J. Text. Inst.* **2012**, *103*, 1255–1261. [CrossRef]
35. Zhou, H.; Jia, B.; Huang, H.; Mou, Y. Experimental Study on Basic Mechanical Properties of Basalt Fiber Reinforced Concrete. *Materials* **2020**, *13*, 1362. [CrossRef]
36. Stephens, D. Natural fiber reinforced concrete blocks. In Proceedings of the 20th WEDC Conference Affordable Water Supply and Sanitation, Colombo, Sri Lanka, 22–26 August 1994; pp. 317–321.
37. Mishra, R.; Behera, B.; Bhagti, V. Influence of noncellulosic contents on nano scale refinement of waste jute fibers for reinforcement in polylactic acid films. *Fiber Polym.* **2014**, *15*, 1500–1506. [CrossRef]
38. Feng, J.; Sun, W.; Zhai, H.; Wang, L.; Dong, H.; Wu, Q. Experimental study on hybrid effect evaluation of fiber reinforced concrete subjected to drop weight impacts. *Materials* **2018**, *11*, 2563. [CrossRef]
39. Marar, K.; Eren, O.; Çelik, T. Relationship between impact energy and compression toughness energy of high-strength fiber-reinforced concrete. *Mater. Lett.* **2001**, *47*, 297–304. [CrossRef]
40. Mishra, R. FEM based prediction of 3D woven fabric reinforced concrete under mechanical load. *J. Build. Eng.* **2018**, *18*, 95–106. [CrossRef]
41. *ASTM C1399/C1399M-10*; Standard Test Method for Obtaining Average Residual-Strength of Fiber-Reinforced Concrete. ASTM international: West Conshohocken, PA, USA, 2015.
42. Alyousef, R.; Ali, B.; Mohammed, A.; Kurda, R.; Alabduljabbar, H.; Riaz, S. Evaluation of Mechanical and Permeability Characteristics of Microfiber-Reinforced Recycled Aggregate Concrete with Different Potential Waste Mineral Admixtures. *Materials* **2021**, *14*, 5933. [CrossRef]
43. Raza, S.; Amir, M.; Azab, M.; Ali, B.; Abdallah, M.; Ouni, M.; Elhag, A. Effect of micro-silica on the physical, tensile, and load-deflection characteristics of micro fiber-reinforced high-performance concrete (HPC). *Case Stud. Constr. Mater.* **2022**, *17*, e01380. [CrossRef]
44. Khan, M.; Cao, M.; Ali, M. Effect of basalt fibers on mechanical properties of calcium carbonate whisker-steel fiber reinforced concrete. *Constr. Build. Mater.* **2018**, *192*, 742–753. [CrossRef]
45. Khan, M.; Cao, M.; Xie, C.; Ali, M. Hybrid fiber concrete with different basalt fiber length and content. *Struct. Concr.* **2022**, *23*, 346–364. [CrossRef]

polymers

MDPI

Article

Mechanical and Thermal Properties of Wood-Fiber-Based All-Cellulose Composites and Cellulose-Polypropylene Biocomposites

Eija-Katriina Uusi-Tarkka [1,*], Mikael Skrifvars [2], Pooria Khalili [2], Henrik Heräjärvi [3], Nawar Kadi [4] and Antti Haapala [1,5]

1 School of Forest Sciences, Faculty of Science and Forestry, University of Eastern Finland, FI-80101 Joensuu, Finland
2 Swedish Centre for Resource Recovery, Faculty of Textiles, Engineering and Business, University of Borås, SE-50190 Borås, Sweden
3 Natural Resources Institute Finland, FI-80100 Joensuu, Finland
4 Department of Textile Technology, Faculty of Textiles, Engineering and Business, University of Borås, SE-50190 Borås, Sweden
5 FSCN Research Centre, Mid Sweden University, SE-85170 Sundsvall, Sweden
* Correspondence: eija.uusi-tarkka@uef.fi

Abstract: This article explores wood-fiber-based fabrics containing Lyocell yarn in the warp and Spinnova–Lyocell (60%/40%) yarn in the weft, which are used to form unidirectional all-cellulose composites (ACC) through partial dilution in a NaOH–urea solution. The aim is to investigate the role of the yarn orientation in composites, which was conducted by measuring the tensile properties in both the 0° and 90° directions. As a reference, thermoplastic biocomposites were prepared from the same fabrics, with biobased polypropylene (PP) as the matrix. We also compared the mechanical and thermal properties of the ACC and PP biocomposites. The following experiments were carried out: tensile test, TGA, DSC, DMA, water absorption test and SEM. The study found no significant difference in tensile strength regarding the Spinnova–Lyocell orientation between ACC and PP biocomposites, while the composite tensile strength was clearly higher in the warp (Lyocell) direction for both composite variants. Elongation at break doubled in ACC in the Lyocell direction compared with the other samples. Thermal analysis showed that mass reduction started at a lower temperature for ACC, but the thermal stability was higher compared with the PP biocomposites. Maximum thermal degradation temperature was measured as being 352 °C for ACC and 466 °C for neat PP, and the PP biocomposites had two peaks in the same temperature range (340–474 °C) as ACC and neat PP combined. ACCs absorbed 93% of their own dry weight in water in just one hour, whereas the PP biocomposites BC2 and BC4 absorbed only 10% and 6%, respectively. The study highlights the different properties of ACC and PP reference biocomposites that could lead to further development and research of commercial applications for ACC.

Keywords: ACC; laminate; mechanical performance; NaOH–urea solvent; single-polymer composite; sustainability; textile structures; thermal analysis; wood fibers

1. Introduction

Environmental concerns, such as dependence on petroleum-based materials, increasing amounts of nonbiodegradable waste and pollution, are among the growing concerns that have driven new biomaterials development [1]. Substituting petroleum-based polymers with natural fibers or other organic sources has resulted in an increasing interest in producing various composites from renewable, nonfossil resources. Cellulose-reinforced composites can be an environmentally friendly option because they originate from renewable resources and have good biodegradability, have low density, are cost effective, have good mechanical performance and are abundant [2,3].

As a semicrystalline polymer, cellulose is difficult to dissolve [4]. During heating, its crystalline phase is stabilized by strong hydrogen bonds, so the cellulose does not go through crystal melting [5]. Solvents that can disturb the intramolecular bonding of cellulose can offer a route to process it into useful products, such as regenerated fibers. Currently, two industrially dominant technologies for creating regenerated cellulose fibers are the Viscose and Lyocell processes [5]. Regenerated fibers are characterized by their high purity, uniformity and reproducibility of mechanical and thermal properties [6]. There are also numerous new types of wood fibers under development and in the early stages of commercial production for the textile industry, where the focus is on improving production efficiency, waste management and recycling, reducing the use of chemicals and water and creating closed-loop production [7,8]. One new generation wood fiber is the Spinnova fiber, which is mechanically refined without dissolution or regeneration from microfibrillated cellulose that originates from raw wood pulp. Spinnova fiber is often combined with other fibers, such as Lyocell, when it is woven into a textile material. Spinnova–Lyocell textile reinforcements could be used in composite manufacturing in the same way as other cellulosic reinforcements, but because it is a new fiber product, limited academic research has yet been conducted.

The interaction between the reinforcement and matrix is crucial for the mechanical and thermal properties of the composites. Thermoplastic polymers, such as polypropylene, are not fully compatible with highly hydrophilic cellulose because of cellulose's abundant hydroxyl groups [9]. The weak interface between the reinforcement and matrix can be improved by physical methods that change the structural and surface properties of the fiber (e.g., surface fibrillation and electric discharge) or by chemical methods (e.g., pretreatments, coupling agents and graft copolymerization) [10,11]. However, treatments usually add costs and can increase chemical intake, as well as creating harmful pollution [12]. An alternative solution to enhance adhesion between the components is to create a single-polymer composite, in which both the reinforcement and matrix are formed by using the same polymer [13,14].

All-cellulose composites (ACCs) are single-polymer composites formed of any type of cellulose. Biodegradable ACCs can be produced by using two different cellulosic sources, which are responsible for forming both the reinforcement and matrix. Another option is to partly dissolve cellulose fibers so that the dissolved cellulose will form a matrix around the reinforcement when the composite is processed by compression and heating. Diverse selections of ACCs can be produced using different methods, solvents and raw materials [8,15]. Nevertheless, less research has been published so far on comparing properties of textile-based materials fabricated either as ACCs or as traditional biocomposites. Finding out how the properties of ACCs stand compared with biocomposites is highly important for further development of the ACCs and for their possibilities in the application range.

The aim and novelty of the present study was to investigate the performance of Lyocell and less-known Spinnova fibers in ACC materials compared with polypropylene matrix composites and to identify the advantages and disadvantages that ACC materials can yield. Because it is known that the direction of the fiber vastly contributes to the properties of the composite [16], unidirectional laminates were examined with a tensile tester for two directions (0° and 90°) to receive a direct comparison of the tensile behavior in the ACC and biocomposites. Thermal analysis and water absorption tests were implemented to gain further knowledge of the composite's performance under changing conditions, which can also greatly affect the material's demand.

2. Materials and Methods

2.1. Materials and Treatments

The textile fabric used in the present study was a plain weave with two-ply Lyocell yarn with 12.6 ends/cm in the warp and 60/40% Spinnova–Lyocell yarn with 16.5 picks/cm in the weft with a surface weight of 173 g/m^2. The material was provided by Spinnova

Ltd., (Jyväskylä, Finland). The structure of the used textile fabric and the corresponding obtained ACC composite is shown in Figure 1.

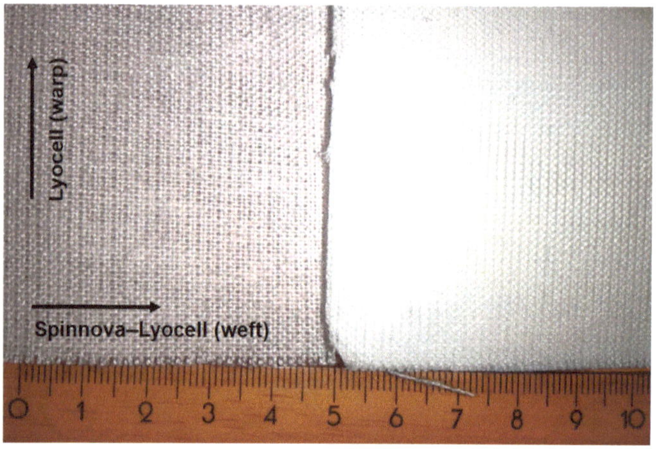

Figure 1. Initial material: Spinnova-Lyocell is seen on the left and produced ACC on the right.

An aqueous NaOH–urea solvent was used to partially dissolve the cellulosic fiber structures. The solvent was prepared by dissolving 12 wt.-% NaOH and 7 wt.-% urea in water at room temperature. The solution was then cooled down to −12 °C and used to prepare the composites by soaking the assembled fabrics in a volume of 0.6 liters. The same solvent batch was used to produce three to four laminate samples. The chemicals were of reagent grade and were supplied by Sigma-Aldrich (Darmstadt, Germany). A biobased (up to 30%) polypropylene was supplied by Nature Plast (Caen, France) and was used to produce the reference PP biocomposites.

2.2. Composite Laminate Preparation

The all-cellulose composite laminates were produced by layering four sheets of fabrics (18×18 cm^2) unidirectionally as a stack. The unidirectional yarn orientation was controlled by a mark on the individual layers. The piles were fully impregnated with the aqueous NaOH–urea solvent for 1 min. Each laminate pile rapidly absorbed around 150 g of the solvent, which was determined by weighing the fabric pile before and after impregnation. The impregnated fabric piles were then placed in 10 L buckets and carefully rinsed with water to remove excess solvent. After rinsing, the samples were left in the buckets in fresh water. Water was changed repeatedly until the pH of the surface of the samples became neutral, as tested using indicator paper. Neutralization took approximately 48 h. If the stacked layers disintegrated during the dissolution or rinsing process, they were manually joined together by tweezers before further processing.

The partly dissolved fabrics were converted to solid composites by pressing the fabric layers together in a hydraulic hot press, where the temperature was set to 100 °C and pressure to 60 bar. The pressing cycles are presented in Table 1. The samples were first pressed for 5 s four times to remove excess water. This was followed by applying 1 min pressure three times, with a 10 s steam release period between the pressings. To continue processing, the samples were pressed once for 3 min and twice for 5 min, here with a 20 s steam release period between the pressings. Finally, the samples were placed in the press for 20 min at 100 °C. After the hot press procedure, the ACC composites (average thickness of 0.60 mm) were wrapped in tissue paper and left between metal plates (fixed hand-tight with screws) overnight at room temperature.

Table 1. Pressing cycles in composite manufacturing for the hot press process.

Cycle Order	Pressure Time	Release Time	Number of Cycles	Pressure	Temperature
1	5 s	10 s	4	60 bar	100 °C
2	1 min	10 s	3	60 bar	100 °C
3	3 min	20 s	1	60 bar	100 °C
4	5 min	20 s	2	60 bar	100 °C
5	20 min	20 s	1	60 bar	100 °C

Thermoplastic biocomposite reference laminates were manufactured by combining the same Spinnova–Lyocell textile with polypropylene (PP). PP granulates were transformed to PP films with the hot press (200 °C), giving films with an approximate thickness of 0.25–0.40 mm. Thinner biocomposites (with average thickness of 0.68 mm) had two layers of cellulose fabric and two layers of PP films (denoted as BC2), and thicker biocomposites (with average thickness of 1.29 mm) had four layers of cellulose fabric and four layers of PP film (denoted as BC4). Each laminate had approximately 45 wt.-% of cellulosic textile fabric and 55 wt.-% of PP. The biocomposites were composed of alternating layers of cellulose fabrics and PP films. The compression molding was carried out in a hot press at 200 °C for 2 min. The obtained biocomposite laminates were finally placed at room temperature between the metal plates for at least 15 min to allow for complete solidification of the PP matrix.

2.3. Composite Characterisation

2.3.1. Tensile Test

Tensile tests were carried out in accordance with the ISO 527 standard using a Tinius Olsen H10KT (Horsham, PE, USA) testing machine. A mechanical extensometer attached to the specimen was used to measure the strain. The rate of loading was 10 mm/min and the load tension was 2.5 kN. The gauge length was 50 mm, and the initial distance between grips was 115 mm. The dumbbell-shaped specimens were cut from the laminates using a laser cutting machine Laser Pro, GLS (Taipei, Taiwan). Tensile tests were performed in two directions (0° and 90°) to determine how fiber orientation affects the mechanical properties of the composites. Twelve replicate specimens were tested in both yarn directions and for all laminate types.

2.3.2. Thermogravimetric Analysis

Thermogravimetric analysis (TGA) and differential thermogravimetric analysis (DTGA) were conducted by an SDT Q500 thermal analyzer (TA instrument, New Castle, DE, USA). Samples of approximately 12 mg were heated at 10 °C/min from 30 °C to 650 °C in a nitrogen environment. The nitrogen flow rate was set to 40 mL/min.

2.3.3. Differential Scanning Calorimetry

Differential scanning calorimetry (DSC) was carried out using the DSC Q1000 instrument (TA instruments, New Castle, DE, USA). An approximately 7 mg sample was heated in a nitrogen purge steam at a heating rate of 10 °C/min from 0 °C to 200 °C, which was followed by cooling and a new heating cycle to determine the glass transition temperature.

2.3.4. Dynamic Mechanical Analysis

Dynamic mechanical analysis (DMA) was performed using a single cantilever test with a Rheometrics Solids Analyzer RSA II (TA Instruments, New Castle, DE, USA). The temperature range was set from 25 °C to 220 °C (note 160 °C for PP). The heating rate was set to 3 °C/min, and the testing frequency was 1 Hz. Test samples were cut to a rectangle shape (50 mm × 10 mm).

2.3.5. Water Absorption Test

A water absorption test was carried out on samples of size 10 × 80 mm^2. The samples were first dried in an oven for 24 h at 60 °C, and then their mass was measured (denoted as W_0). During the first day, two additional measurements were taken: weight after 1 h and 3 h of water exposure. The next measurements were scheduled after 24 h for the next 6 days. The mass at each measurement point in time was compared with W_0 to detect the amount of water absorbed as a function of time. The mass gain (m) is the percentage change in mass relative to the initial mass, which was calculated using Equation (1). The average mass gain was calculated from five replicated samples.

$$m = \frac{W - W_0}{W_0} \cdot 100\% \quad (1)$$

2.3.6. Scanning Electron Microscopy

Morphological changes in the cross-section area were detected using scanning electron microscope (SEM) imaging carried out using a Hitachi S-4800 (Hitachi, Tokyo, Japan). The samples were gold-coated (2 nm) with a Cressington sputter coater 208 HR (Watford, UK) before the imaging.

3. Results and Discussion

3.1. Tensile Test

Tensile tests were carried out in two directions, where S stands for the Spinnova–Lyocell (weft) direction and L stands for the Lyocell (warp) direction. The thinner polypropylene reference biocomposites were marked as S-BC2 and L-BC2, and the thicker polypropylene biocomposites were marked as S-BC4 and L-BC4. ACCs were marked as S-ACC and L-ACC, depending on the direction of the test.

The tensile strength (MPa) results show (Figure 2) that there was practically no difference between the ACC and PP biocomposites, BC2 and BC4, measured in the Spinnova–Lyocell direction; indeed, the ACCs and biocomposites had quite similar strengths (35 +/− 2 MPa). All the composites had higher tensile strength in the Lyocell direction compared with the Spinnova–Lyocell direction, indicating better mechanical strength for the Lyocell fibers. This could be explained by the Spinnova fiber being more dissolved in the NaOH–urea dissolution process than the Lyocell fiber. The composite will therefore have better strength in the Lyocell direction. The highest tensile strength, 51 MPa, was measured for the L-BC4, and the lowest, 43 MPa, for the ACC in the Lyocell direction.

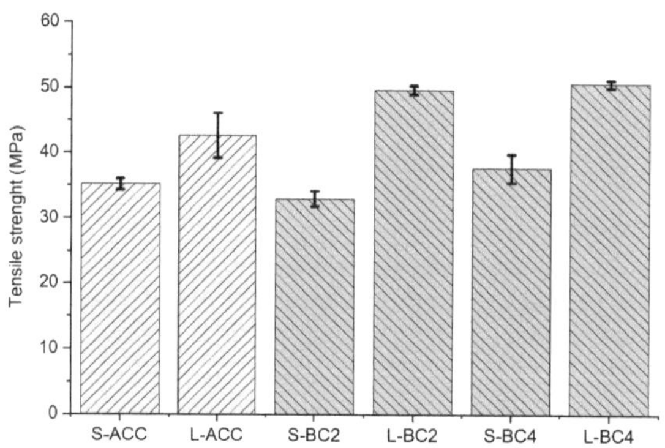

Figure 2. Tensile strength (MPa) for ACC and PP reference biocomposites. Tensile strength measurements were performed in two directions, where S stands for the Spinnova–Lyocell (weft) direction and L stands for the Lyocell (warp) direction. Lighter shade is used for ACC, darker for BC.

A study comparing ACCs and biocomposites was published by Gildl-Altmutter et al. [17], who reported a tensile strength of 34 MPa for nonwoven flax-based ACC, which is comparable to the 35 MPa value for S-ACC seen in our experiment. However, in their study, the tensile strength for flax–epoxy biocomposites was measured as being 79 MPa, which is more than double compared with their flax-based ACC. In our study, the difference between ACC and reference PP biocomposites in the Lyocell direction was only 2 MPa for the Spinnova–Lyocell direction and 7–8 MPa for the Lyocell direction. Nevertheless, comparing our results to other studies is complicated because of differences in the amounts of fibers, their purities, solvents used and manufacturing techniques and conditions.

Similarly, the elongation at break (%) did not vary much between the S-ACC, S-BC2 and S-BC4 (Figure 3) specimens. The results also indicate that elongation at break (%) was higher for all composites in the Lyocell direction than in the Spinnova–Lyocell direction. The number of fabric layers, either two or four, in the PP biocomposites did not greatly affect the properties. One interesting result was that elongation at break (%) value for L-ACC was as high as 38%, whereas other composites' received values were between 13% and 21%. Adak and Mukhopadhyay [18] measured the tensile properties of Lyocell-fabric-based ACC using eight textile layers and ionic liquids as solvents. For a half an hour dissolution time, they received elongation at break values of roughly 26% in the warp direction and 30% in the weft direction. The elongation at break values in our L-ACC specimens were higher than those in the specimens of Adak and Mukhopadhyay [18]. A reason for this might be that Lyocell fiber, which has a skin–core structure [19], is more elastic than the Spinnova fiber, which is produced without regeneration [20]. In addition, L-ACC does not have a cohesive matrix, such as polypropylene in L-BC2 and L-BC4, which will keep the fiber in place and prevent greater elongation. Different textile layers are also strongly attached to each other in the PP biocomposites compared with ACC. This is visualized in the broken tensile test samples (Figure 4).

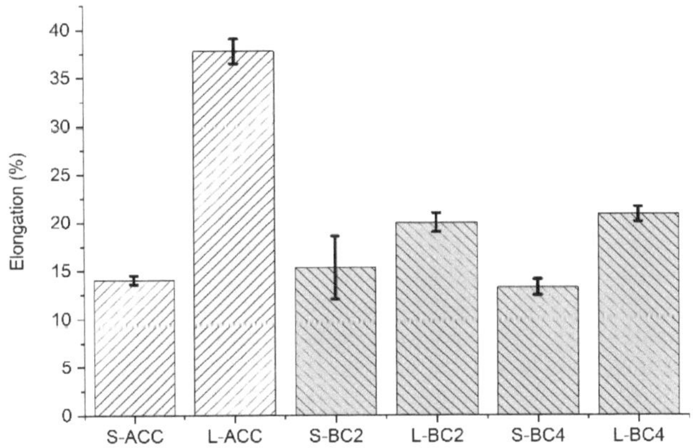

Figure 3. Elongation at break (%) for ACC and biocomposites. Measurements were performed in two directions, where S stands for the Spinnova–Lyocell (weft) direction and L stands for the Lyocell (warp) direction. Lighter shade is used for ACC, darker for BC.

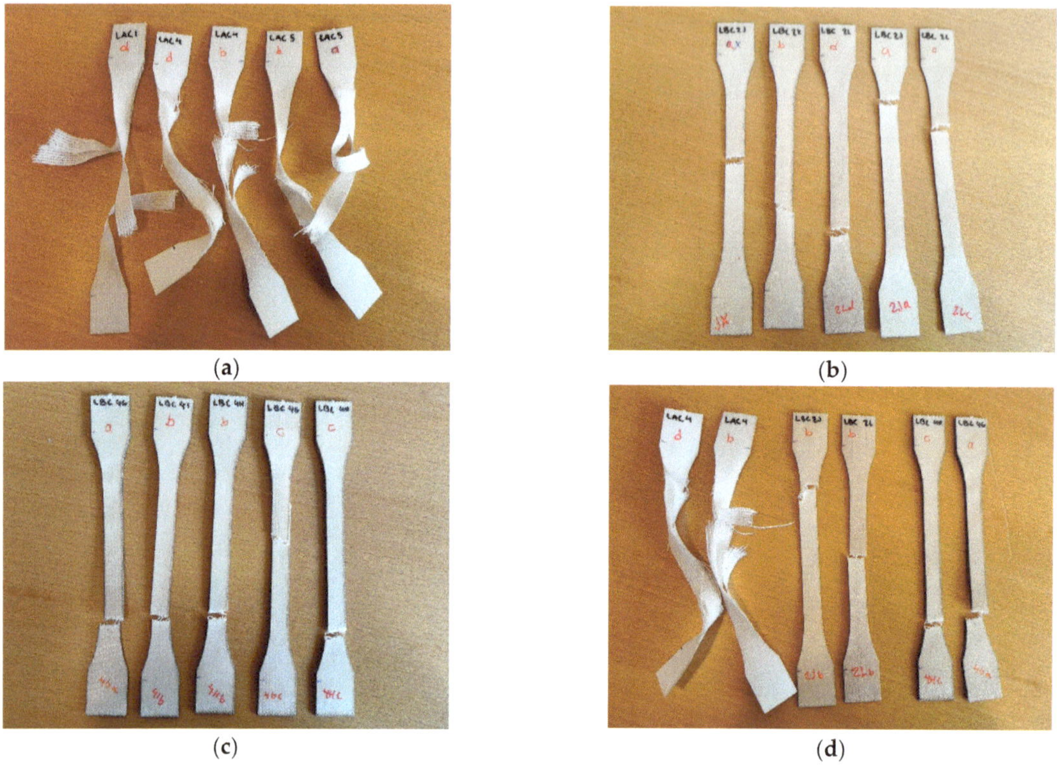

Figure 4. Broken dumbbell tensile test samples. Test samples of L-ACC are seen in (**a**), L-BC2 in (**b**) and L-BC4 in (**c**), and two of each sample category in the same order are seen in image (**d**).

3.2. Thermogravimetric Analysis

Thermogravimetric analysis (TGA) quantified the thermal resistance by recording the percentage weight loss of the samples when there were increasing temperatures. In the thermal analyses, there was no difference between the Lyocell and Spinnova–Lyocell yarn directions; therefore, only each type of composite, that is, ACC, BC2 and BC4, were examined. In addition, neat PP film was tested to give a reference value for the PP biocomposites.

The TGA results (Figure 5) show that there was a small mass reduction for ACC during the beginning of the heating cycle, which was because of evaporating water. This was also reported in previous studies [21–23]. A further sharp decrease in mass reduction for ACC started at around 310 °C, and a steadier decrease continued at around 360 °C. For PP biocomposites, a sharp mass reduction started later at around 420 °C. It also approached zero in a single step without leaving any significant residue. The PP biocomposites behaved almost identically in TGA, and we could see a clear difference between them and the ACC and neat PP. It was also visible that degradation decreased in the two different temperature regions where the curves were corresponding values of ACC and PP. The biocomposites' final mass residue was around 10% in the end; therefore, compared with PP, it had higher thermal stability, while ACC had the highest.

Figure 6 shows the results of the differential thermogravimetric analysis (DTGA). The maximum thermal degradation temperature was measured at 352 °C for ACC and 466 °C for PP biocomposites. For the biocomposites, thermal degradation occurred in a similar temperature range (from 340 °C to 474 °C) than ACC and PP. The first peaks indicating maximum thermal degradation were attributed to the cellulose reinforcement, and the second peaks were attributed to the PP polymer matrix. The maximum decomposition rates were close to the ACC and PP values, but there was a small shift where the high-intensity areas were located, which could be because of the changes in mass. The values of the first peaks for BC2 and BC4 were 339 °C and 340 °C and, for the second peaks, 469 °C and 475 °C, respectively.

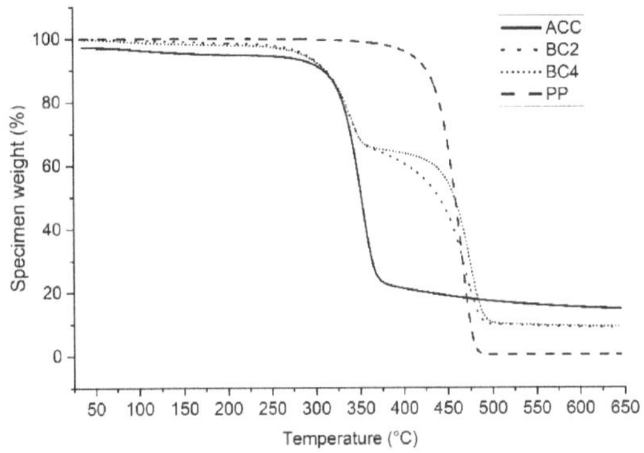

Figure 5. Results of TGA for all-cellulose composites (ACC), polypropylene biocomposites (BC2 and BC4) and neat polypropylene (PP).

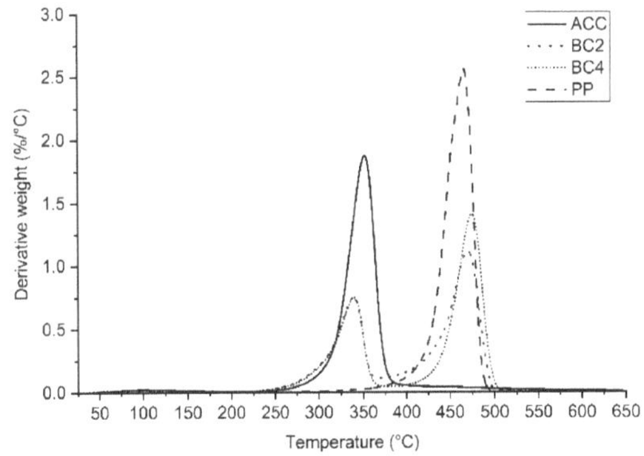

Figure 6. Results of the DTGA for all-cellulose composites (ACC), biocomposites (BC2 and BC4) and polypropylene (PP).

Measured values from thermal testing depend on parameters such as nitrogen rate, sample weight and heating rate [24]. In a nitrogen environment, the thermal decomposition of PP has previously been recorded to start at 300 °C and end at 475 °C [25], which is ~10 °C lower than what was recorded in our study. Yang et al. [22] manufactured ACCs from regenerated fibers and reported that the initial decomposition temperatures of modified ramie fibers, natural ramie fibers and regenerated cellulose films were 281 °C, 262 °C and 270 °C, respectively, while the peak of the modified ramie fibers was approximately 356 °C. The results indicate that ACCs made of Spinnova–Lyocell cellulose fibers in our research have as high thermostability as modified ramie-fiber-based ACCs.

3.3. *Differential Scanning Calorimetry*

Differential scanning calorimetry (DSC) can detect any residual exothermal heat present in the samples. As shown in Figure 7, the exothermic peak, which represents crystallization temperature (T_c), was measured at 122 °C for PP, BC4 and BC2, while the respective heat flows were 2.6 W/g, 1.5 W/g and 1.4 W/g. The melting temperatu©T_m for the PP biocomposites and neat PP was measured at 164 °C. Similar results have been reported in the literature: T_m was detected at 165 °C for pure PP (Ibrahim et al., 2016) and 163 °C for biocomposites modeled using microcrystalline cellulose and PP/PLA [26]. The main difference between PP, PP biocomposites and ACC was that ACC did not have an explicit T_c or T_m, as expected. Cellulose has a strong network of hydrogen bonds and undergoes thermal decomposition when heated [27,28].

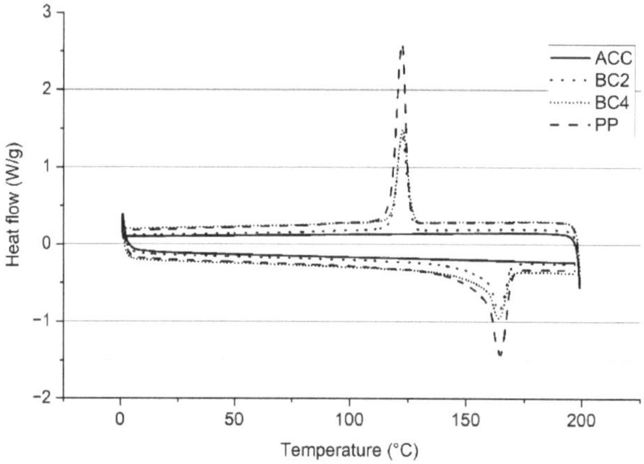

Figure 7. Results of DSC in a range between 0 °C and 200 °C for all-cellulose composites (ACC), biocomposites (BC2 and BC4) and polypropylene (PP).

3.4. *Dynamic Mechanical Analysis*

The dynamical mechanical analysis (DMA) measures the viscoelastic properties of the samples. The results are divided into three categories: storage modulus (GPa) (Figure 8a), loss modulus (GPa) (Figure 8b) and tan delta (δ) (Figure 8c). According to DSC, the melting temperature for PP was 164 °C. Therefore, the analysis was programmed to end at 150 °C for PP, while it lasted longer (~220 °C) for the other samples.

The storage modulus represents the energy stored in the elastic structure of the sample (Frank, TA instruments). During the beginning of the test (25 °C), the thickest PP biocomposite with four layers of fabric (BC4) had the highest storage modulus (1.9 GPa), followed by BC2 (1.7 GPa), ACC (1.3 GPa) and pP (1 GPa). The PP and reference PP biocomposites had a steady decrease in storage modulus until 150 °C. When these samples approached the melting temperature of PP, the storage modulus remained at zero. ACC lost less than 40% of its storage modulus during the experiment (from 1.3 GPa to 0.8 GPa). This phenomenon, where ACCs' storage modulus remained high and relatively steady, has also been reported previously by Soykeabkaew and Peijis [29]. According to the results, the viscoelastic properties of PP biocomposites BC2 and BC4 did not resemble the properties of ACC but rather the properties of neat PP, which was expected. The structure of the biocomposites and their ability to elastically respond to stress was lost when PP reached near T_m.

The loss modulus represents the viscous part, or the amount of energy dissipated in the sample [30]. Evaporating water in ACC created a small reduction in the loss modulus, but overall, the values stayed high when compared with neat PP and PP biocomposites. Thermal softening and a decrease in loss modulus started to gradually occur for the neat PP and PP biocomposites at 60 °C, which was followed by a sharp decrease at 140 °C for the PP biocomposites. BC4 had a higher amount of cellulose fabric but also a higher amount of PP, which can be seen as a lower loss modulus after 150 °C compared with BC2. The behavior of the PP matrix influenced the thermal properties of biocomposites more strongly than its cellulosic filler.

Tan delta is the ratio between the loss modulus and storage modulus. The PP and PP biocomposite results are not meaningful to compare after PP has reached its limits regarding viscoelastic stress, since once this deformation of the composites has taken place, the material will not recover to its original form. The glass transition temperature (T_g) for polypropylene varies depending on its crystallinity, but it is recorded to be below −25 °C [31], around 0 °C [32] or just above −12 °C [33]. In this graph, T_g is not detectable for neat PP, PP biocomposites, or ACC. Even though cellulose underwent a second-order phase transition from the low-temperature glass state to the high-temperature elastic state [34,35], earlier research has reported that there was not necessarily a clear T_g for cellulose [17,36], as was the case here.

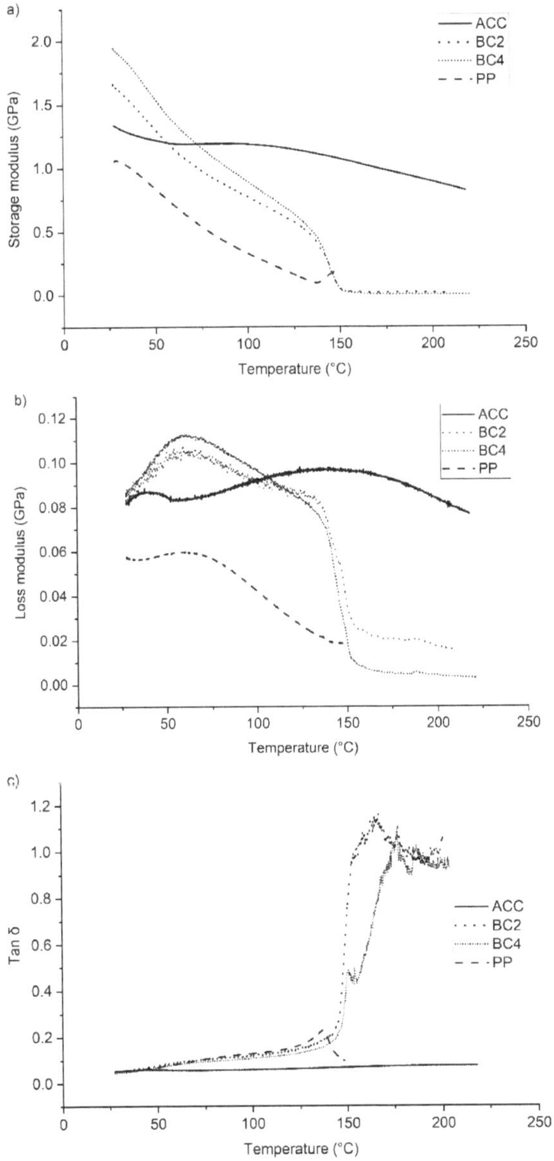

Figure 8. Results of the DMA: (**a**) storage modulus (GPa), (**b**) loss modulus (GPa) and (**c**) tan δ for all-cellulose composites (ACC), biocomposites (BC2 and BC4) and polypropylene (PP).

3.5. Water Absorption Test

Figure 9 shows the weight gain (%) of the composites as a function of water immersion time. The graph shows that all composites absorbed some water after immersion. However, there was a major difference in the water absorption rate between ACC and PP biocomposites already after 1 h. During this first hour, ACC absorbed 93% of its own dry weight in water, whereas BC2 and BC4 absorbed only 10% and 6%, respectively. After a day, ACC reached 97%, meaning that composites almost doubled their weight in 24 h. On the fourth day, all the composites had absorbed close to their maximum capacity of water, and the absorption rate leveled off during the following days. The maximum water absorption values were reached on day 5 for ACC (117%) and BC4 (30%) and on day 6 for the BC2 (35%). There was no considerable difference between BC2 and BC4, but water intake was slightly lower for BC4. The share of hydrophobic PP matrix was higher in BC4, which likely contributed to the composites' low water absorption rate. Hydrophilic ACCs' tendency to absorb water will greatly limit their possible applications without a waterproof coating or modification.

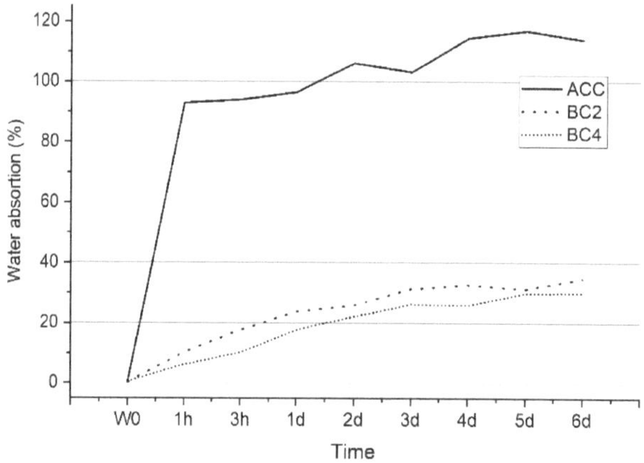

Figure 9. Results from a water absorption test of 6 days, where W_0 stands for dry weight.

3.6. Scanning Electron Microscopy

Scanning electron microscopy (SEM) pictures were taken from the neat Spinnova–Lyocell fabric (Figure 10). The different fiber structures and width of the yarns are visible in the images. Cross-sectional SEM pictures of S-ACC, L-ACC, S-BC2, L-BC2, S-BC4 and L-BC4 composites can be seen in Figure 11, here with a magnification of 80×. Figure 12 presents the S-ACC, L-ACC, S-BC2 and L-BC2 structures with a magnification of 180×. All images to the left (Figures 11a,c,e and 12a,c) are composite cross-sections with the Spinnova–Lyocell fibers in the horizontal direction (out of plane) and the Lyocell fibers in the vertical direction. All images on the right side (Figures 11b,d,f and 12a,c) have the fibers in opposite direction. A clear difference can be seen between images (a) S-ACC and (b) L-ACC in Figures 10 and 11. During composite processing, the Spinnova fiber dissolved in greater extent compared with the Lyocell fiber. In the image of S-ACC, (a) the Spinnova fibers are well fused together, but in L-ACC (b) individual undissolved Lyocell fibers are still visible as independent fibers. In both ACC images, different layers of textile fabrics are clearly visible. In the case of the reference PP biocomposites (In Figures 11c–f and 12c,d), melted PP has shaped the structure of the composite so that cellulosic fibers or individual layers of the textile cannot be clearly detected anymore. This is even more evident in the thicker composites S-BC4 (Figure 11e) and L-BC4 (Figure 12f).

Figure 10. SEM pictures of Spinnova–Lyocell fabric with Lyocell yarn in the warp (longitudinal direction) and 60/40% Spinnova–Lyocell yarn in the weft (horizontal direction), with magnifications of (**a**) 30× and (**b**) 80×.

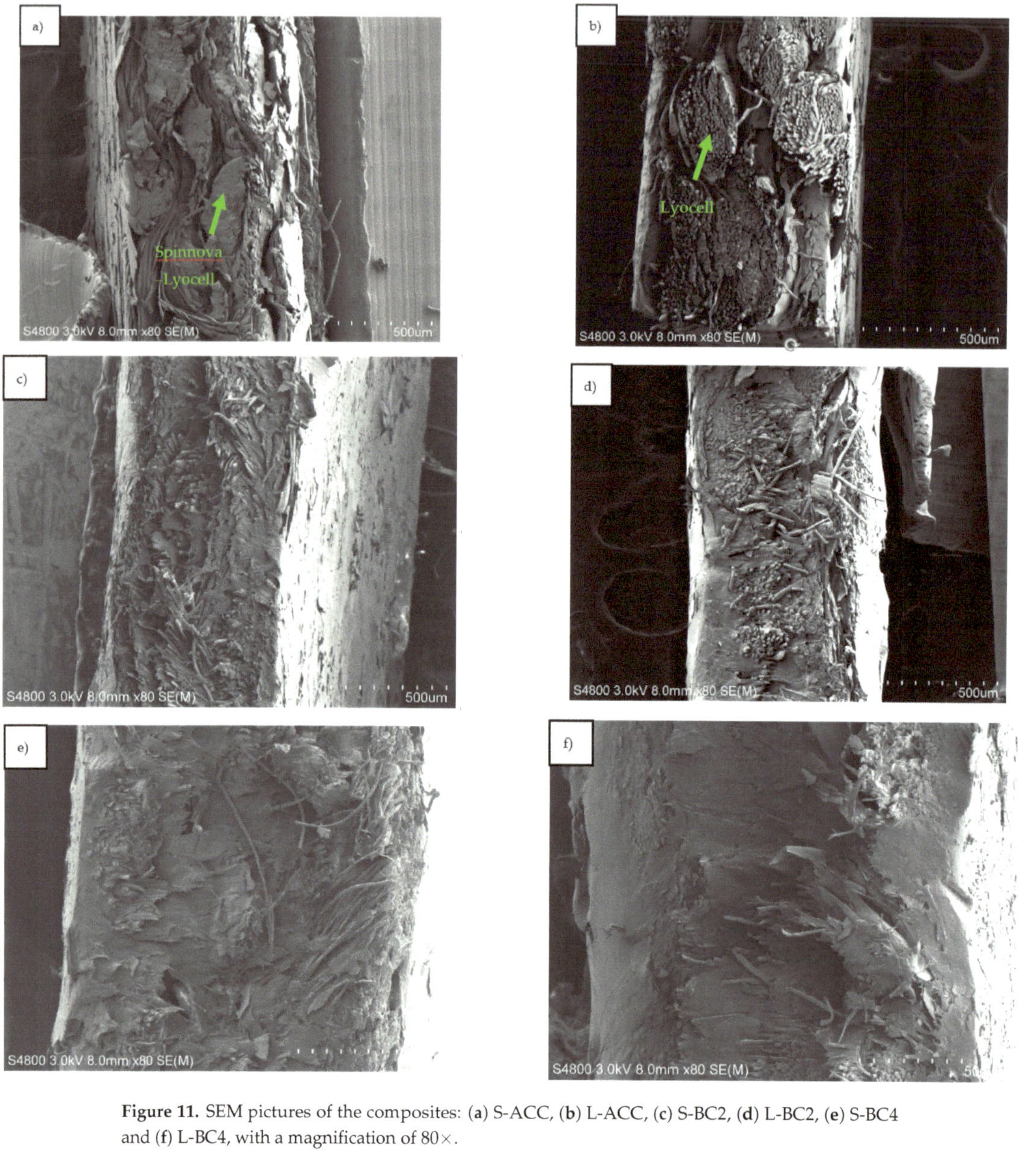

Figure 11. SEM pictures of the composites: (**a**) S-ACC, (**b**) L-ACC, (**c**) S-BC2, (**d**) L-BC2, (**e**) S-BC4 and (**f**) L-BC4, with a magnification of 80×.

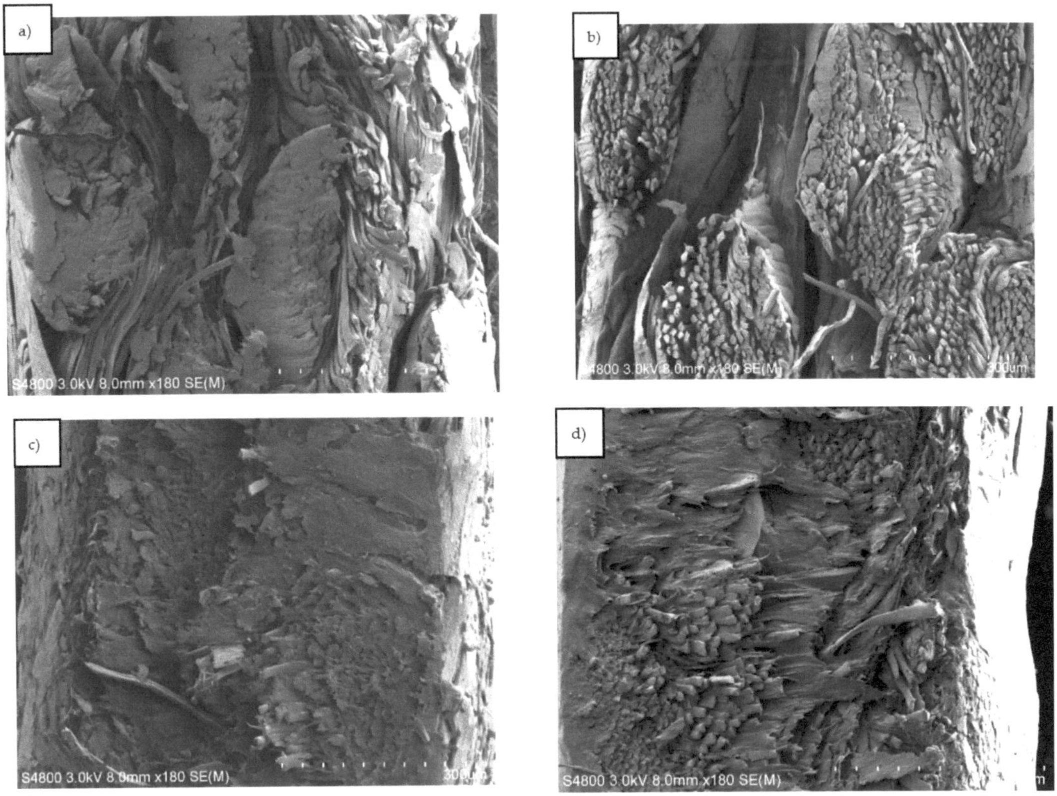

Figure 12. SEM pictures of the composites: (**a**) S-ACC, (**b**) L-ACC, (**c**) S-BC2 and (**d**) L-BC2, with a magnification of 180×.

4. Conclusions

The current study revealed that the direction of the studied yarns greatly contributed to the composites' tensile strength and elongation properties. There was no considerable difference in the tensile strength between the ACC and PP reference biocomposites when measured in the Spinnova–Lyocell yarn direction, whereas when measured in the Lyocell yarn direction, the tensile strength was higher in the PP reference biocomposites than in the ACCs. Overall, the Lyocell-reinforced composites had better tensile strength compared with the Spinnova–Lyocell-reinforced composites. The main finding in elongation studies was that the L-ACC composite has a remarkably higher elongation at break (38%) compared with any of the other composites. Regenerated Lyocell seems to have very textile-like characteristics, even as ACC, whereas in the reference biocomposites, the stiffness of the PP matrix prevented greater elongation. In addition, the individual reinforcement layers in the ACC were not as well bonded to each other as they were in the reference PP biocomposites. A challenge in the development of ACCs is therefore to prevent the unwanted delamination of the textile layers. With PP biocomposites, the orientation of the fibers had more impact on the properties of the composite than the number of polymer layers used.

The thermal studies demonstrated that ACCs had a better thermal stability than the PP biocomposites. PP dominated the thermal behavior in the reference biocomposites, which was expected. T_c and T_m values of 122 °C and 164 °C, respectively, were detected for the PP biocomposites, whereas no crystallization or melting temperatures were identified for cellulose composites. The high cross-linking of cellulose restricted the movements of the chain segments during heating, which would be required for fluid transformation. Instead, the cellulose degrades at high temperatures and does not show any thermal transitions. The water absorption test showed the highly hydrophilic nature of cellulose: ACC absorbed twice its weight in water in the immersion test in less than 48 h. Because of the protective layer of PP, the biocomposites resisted wetting much better, and there was only a minor difference between BC2 and BC4 composites.

The SEM images showed clear differences in structure for the two studied cellulose fibers. Lyocell and Spinnova–Lyocell had different fiber structures in neat fabrics, which were also detectable in the composites. In the reference biocomposites, especially in the thicker laminates, the different fabric layers could not be as easily seen in the cross-sectional images as in the ACCs. The reported findings regarding the role of direction of the fiber and properties of ACC compared with PP biocomposites are important for further optimization, product development and use in structural composite applications.

Author Contributions: Conceptualization E.-K.U.-T. and M.S.; methodology E.-K.U.-T., M.S. and P.K.; formal analysis, investigation, resources and writing—original draft preparation E.-K.U.-T.; visualization, data curation, software and validation E.-K.U.-T. and A.H.; writing—review and editing, M.S., P.K., H.H., N.K. and A.H.; supervision M.S., H.H., N.K. and A.H.; project administration and funding acquisition, E.-K.U.-T. and A.H. All authors have read and agreed to the published version of the manuscript.

Funding: The authors thank the Finnish Natural Resources Research Foundation, the Finnish Cultural Foundation and the Magnus Ehrnrooth Foundation for cofunding the project leading to this work.

Institutional Review Board Statement: Not applicable.

Informed Consent Statement: Not applicable.

Data Availability Statement: The data presented in this study are available on request from the corresponding author.

Acknowledgments: We thank Spinnova Ltd. For providing material for our research. Eija-Katriina also wants to thank her coworkers at the University of Borås, and especially the polymer group, which made her stay delightful and helped her achieve her goals.

Conflicts of Interest: The authors declare no conflict of interest. The funders had no role in the design of the study; in the collection, analyses, or interpretation of data; in the writing of the manuscript; or in the decision to publish the results.

References

1. Mohanty, A.K.; Misra, M.; Hinrichsen, G. Biofibres, biodegradable polymers and biocomposites: An overview. *Macromol. Mater. Eng.* **2000**, *276*, 1–24. [CrossRef]
2. Hu, F.; Wang, M.; Wang, N.; Hu, Y.; Gan, M.; Liu, D.; Xie, Y.; Feng, Q. All-cellulose composites prepared by partially dissolving cellulose using NaOH/thiourea aqueous solution. *J. Appl. Polym. Sci.* **2021**, *138*, 51298. [CrossRef]
3. Carrillo, F.; Colom, X.; Cañavate, X. Properties of Regenerated Cellulose Lyocell Fiber-Reinforced Composites. *J. Reinf. Plast. Compos.* **2010**, *29*, 359–371. [CrossRef]
4. Fujisawa, S.; Saito, T.; Isogai, A. All-cellulose (cellulose-cellulose) green composites. In *Advanced Green Composites*; John and Wiley and Sons: Hoboken, NJ, USA, 2018; pp. 111–133. [CrossRef]
5. Sharma, A.; Nagarkar, S.; Thakre, S.; Kumaraswamy, G. Structure–property relations in regenerated cellulose fibers: Comparison of fibers manufactured using viscose and lyocell processes. *Cellulose* **2019**, *26*, 3655–3669. [CrossRef]
6. Lenz, J.; Shurz, J.; Eichinger, D. Properties and Structure of Lyocell and Viscose-Type Fibers in the Swollen State. *Lezinger Ber.* **1994**, *9*, 19–25.
7. Mathur, K. New Developments in Fibers, Yarns & Fabrics. *Text. World* **2020**, *170*, 32–35.
8. Uusi-Tarkka, E.-K.; Skrifvars, M.; Haapala, A. Fabricating sustainable all-cellulose composites. *Appl. Sci.* **2021**, *11*, 10069. [CrossRef]
9. Trejo-O'Reilly, J.A.; Cavaillé, J.Y.; Paillet, M.; Gandini, A.; Herrera-Franco, P.; Cauich, J. Interfacial properties of regenerated cellulose fiber/polystyrene composite materials. Effect of the coupling agent's structure on the micromechanical behavior. *Polym. Compos.* **2000**, *21*, 65–71. [CrossRef]
10. Jayamol, G.; Sreekala, M.S.; Sabu, T. A review on interface modification and characterization of natural fiber reinforced plastic composites. *Polym. Eng. Sci.* **2001**, *41*, 1471–1485. [CrossRef]

11. Zafeiropoulos, N.E. *Interface Engineering of Natural Fibre Composites for Maximum Performance*; Woodhead Publishing: Sawston, UK, 2011.
12. Arévalo, R.; Peijs, T. Binderless all-cellulose fibreboard from microfibrillated lignocellulosic natural fibres. *Compos. Part A Appl. Sci. Manuf.* **2016**, *83*, 38–46. [CrossRef]
13. Alcock, B.; Cabrera, N.O.; Barkoula, N.-M.; Loos, J.; Peijs, T. The mechanical properties of unidirectional all-polypropylene composites. *Compos. Part A Appl. Sci. Manuf.* **2006**, *37*, 716–726. [CrossRef]
14. Capiati, N.J.; Porter, R.S. The concept of one polymer composites modelled with high density polyethylene. *J. Mater. Sci.* **1975**, *10*, 1671–1677. [CrossRef]
15. Baghaei, B.; Skrifvars, M. All-Cellulose Composites: A Review of Recent Studies on Structure, Properties and Applications. *Molecules* **2020**, *25*, 2836. [CrossRef] [PubMed]
16. Ramamoorthy, S.K.; Skrifvars, M.; Persson, A. A Review of Natural Fibers Used in Biocomposites: Plant, Animal and Regenerated Cellulose Fibers. *Polym. Rev.* **2015**, *55*, 107–162. [CrossRef]
17. Gindl-Altmutter, W.; Keckes, J.; Plackner, J.; Liebner, F.; Englund, K.; Laborie, M.-P. All-cellulose composites prepared from flax and lyocell fibres compared to epoxy–matrix composites. *Compos. Sci. Technol.* **2012**, *72*, 1304–1309. [CrossRef]
18. Adak, B.; Mukhopadhyay, S. Effect of the dissolution time on the structure and properties of lyocell-fabric-based all-cellulose composite laminates. *J. Appl. Polym. Sci.* **2016**, *133*, 43398. [CrossRef]
19. Fink, H.P.; Weigel, P.; Purz, H.; Ganster, J. Structure formation of regenerated cellulose materials from NMMO-solutions. *Prog. Polym. Sci.* **2001**, *26*, 1473–1524. [CrossRef]
20. Adusumali, R.-B.; Reifferscheid, M.; Weber, H.; Roeder, T.; Sixta, H.; Gindl, W. Mechanical Properties of Regenerated Cellulose Fibres for Composites. *Macromol. Symp.* **2006**, *244*, 119–125. [CrossRef]
21. Tanpichai, S.; Witayakran, S. All-cellulose composites from pineapple leaf microfibers: Structural, thermal, and mechanical properties. *Polym. Compos.* **2018**, *39*, 895–903. [CrossRef]
22. Yang, Q.; Lue, A.; Zhang, L. Reinforcement of ramie fibers on regenerated cellulose films. *Compos. Sci. Technol.* **2010**, *70*, 2319–2324. [CrossRef]
23. Uusi-Tarkka, E.-K.; Levanic, J.; Herajarvi, H.; Kadi, N.; Skrifvars, M.; Haapala, A. All-Cellulose Composite Laminates Made from Wood-Based Textiles: Effects of Process Conditions and the Addition of TEMPO-Oxidized Nanocellulose. *Polymers* **2022**, *14*, 3959. [CrossRef] [PubMed]
24. Navarro, R.; Torre, L.; Kenny, J.M.; Jiménez, A. Thermal degradation of recycled polypropylene toughened with elastomers. *Polym. Degrad. Stab.* **2003**, *82*, 279–290. [CrossRef]
25. Esmizadeh, E.; Tzoganakis, C.; Mekonnen, T.H. Degradation behavior of polypropylene during reprocessing and its biocomposites: Thermal and oxidative degradation kinetics. *Polymers* **2020**, *12*, 1627. [CrossRef] [PubMed]
26. Bhasney, S.M.; Kumar, A.; Katiyar, V. Microcrystalline cellulose, polylactic acid and polypropylene biocomposites and its morphological, mechanical, thermal and rheological properties. *Compos. Part B Eng.* **2020**, *184*, 107717. [CrossRef]
27. Belgacem, M.N.; Gandini, A. Production, chemistry and properties of cellulose-based materials. In *Biopolymers—New Materials for Sustainable Films and Coatings*; Plackett, D., Ed.; Wiley: West Sussex, UK, 2011; pp. 151–178.
28. Nishino, T.; Matsuda, I.; Hirao, K. All-Cellulose Composite. *Macromolecules* **2004**, *37*, 7683–7687. [CrossRef]
29. Soykeabkaew, N.; Nishino, T.; Peijs, T. All-cellulose composites of regenerated cellulose fibres by surface selective dissolution. *Compos. Part A Appl. Sci. Manuf.* **2009**, *40*, 321–328. [CrossRef]
30. Franck, A. Viscoelasticity and Dynamic Mechanical Testing, TA Instruments. Germany. 2022. Available online: https://www.tainstruments.com/pdf/literature/AAN004_Viscoelasticity_and_DMA.pdf (accessed on 11 November 2022).
31. Greene, J.O. 3—Microstructures of Polymers. In *Plastics Design Library, Automotive Plastics and Composites*; Greene, J.P., Ed.; William Andrew Publishing: Norwich, NY, USA, 2021; pp. 27–37. ISBN 9780128180082. [CrossRef]
32. Woicke, N.; Keuerleber, M.; Hegemann, B.; Eyerer, P. Three-dimensional thermorheological behavior of isotactic polypropylene across glass transition temperature. *J. Appl. Polym. Sci.* **2004**, *94*, 877–880. [CrossRef]
33. Gómez-del Río, T.; Rodríguez, J. Compression yielding of polypropylenes above glass transition temperature. *Eur. Polym. J.* **2010**, *46*, 1244–1250. [CrossRef]
34. Salmén, N.L.; Back, E.L. The influence of water on the glass phase transition temperature of cellulose. *TAPPI* **1977**, *60*, 137–140.
35. Wunderlich, B. *Thermal Analysis of Polymeric Materials*; Springer: Berlin/Heidelberg, Germany, 2005.
36. Vittadini, E.; Dickinson, L.; Chinachoti, P. H-1 and H-2 NMR mobility in cellulose. *Carbohydr. Polym.* **2001**, *46*, 49–57. [CrossRef]

polymers

MDPI

Article

Dimensional Stability of Light-Activated Urethane Dimethacrylate Denture Base Resins

Swati Mishra [1], Saurabh Chaturvedi [2,*], Mariyam Ali [1], Kaushik Kumar Pandey [1], Nasser M. Alqahtani [2], Mohammed A. Alfarsi [2], Mohamed Khaled Addas [2], Sunil Kumar Vaddamanu [3], Nasser M. Al Ahmari [2], Saeed M. Alqahtani [2], Ashfaq Yaqoob [2] and Waleed M. S. Alqahtani [2]

[1] Department of Prosthodontics, Career Postgraduate Institute of Dental Sciences, Lucknow 226008, India
[2] Department of Prosthetic Dentistry, College of Dentistry, King Khalid University, Abha 61421, Saudi Arabia
[3] Department of Dental Technology, College of Applied Medical Sciences, King Khalid University, Abha 61421, Saudi Arabia
* Correspondence: survedi@kku.edu.sa

Abstract: An accurate and dimensionally stable trial denture base is required for a successful denture. The aim of this in vitro study was to assess the dimensional stability of a light-activated urethane dimethacrylate (UDMA) visible light cure (VLC) denture base with three fabrication techniques and different curing cycles. Forty-five VLC denture base samples were divided evenly into three groups. Group A used a conventional fabrication technique with a curing cycle of 5 min. Group B used a modified fabrication technique with two 4-min curing cycles. Group C used a multi-step fabrication technique with three curing cycles (4 min, plus 4 min, plus 2 min). The samples were sectioned and observed under a stereomicroscope to measure the discrepancy between the sample and the master cast. The mean dimensional discrepancy (mm) at the molar region at mid-palate, after 24 h in Group A, B and C was 0.790 mm, 0.741 mm and 0.379 mm, respectively; at the right ridge crest, it was 0.567, 0.408 and 0.185, while at the left ridge crest it was 0.475, 0.331 and 0.125, respectively. Statistical analysis showed significantly different dimensional discrepancies among the groups at all three sites; right ridge crest (F = 93.54, $p < 0.001$), left ridge crest (F = 105.96, $p < 0.001$) and mid-palate (F = 125.53, $p < 0.001$). Within the limitations of this laboratory study, it can be concluded that the denture base using a multi-step fabrication technique with three curing cycles provides better adaptation than the conventional technique. The significance of the study is that clinicians should consider performing denture base fabrication using a multi-step technique to enhance adaptation and hence the stability of the dentures for patients.

Keywords: UDMA; record base; light cure denture base; dimensional discrepancy

1. Introduction

An accurate and dimensionally stable trial denture base is mandatory for the construction of a successful denture. To achieve intraoral stability and retention of the record base, the base should maintain close adaptation to the cast and should be dimensionally stable. The adaptation of a record base, however, depends on many factors, including the materials and methods used for its construction. It is a proven fact that the more dimensionally accurate and stable a material is, the more retentive and stable the denture will be. In the past, various materials were used to fabricate custom trays, record bases and dentures such as Shellac, Thermoform, Polycaproilaitone, Self-Cure Acrylic Resin and Heat Cure Acrylic Resin. The problem with these acrylic resins is that they shrink and become discoloured over time due to the absorption of water and oral fluids. They also undergo significant volumetric and optical changes in the conditions of the oral environment [1–5].

A lack of dimensional stability is one of the major shortcomings of acrylic resins. These dimensional changes occur either in the form of shrinkage or expansion and are mainly affected by the method of curing, along with other factors. The polymerization shrinkage during curing and the stresses released during flask cooling in heat-cured acrylic resins result in changes in dimensions and their magnitude is mainly affected by acrylic resin thickness [6–8]. Other factors include polymerization techniques, in which the difference in thermal expansion of gypsum and acrylic resin results in the development of internal stresses and variation in base thickness at different sites inside the flask results in altered record base adaptation and stability [9–12]. On the other hand, few authors stated that there is a reduction in the molecular weight of the resulting polymer chains after curing and so the variation in the curing techniques may not alter significantly the pattern of dimensional changes in acrylic resin.

The changes in the dimensions of the acrylic resins may affect patient satisfaction and negatively impact the stability and retention of record bases and ultimately the success of the prosthesis [9]. In their study, Consani et al. showed that the molar region is the most consistent site to produce the gap between the palatal area and the acrylic record base. They reasoned that it represents a linear shrinkage which occurs in the acrylic resin [10].

Visible light cure denture has attracted special attention recently as a new record base material and has been used worldwide as an alternative to the chemical acrylic resin material since 1984. It was marketed under the trade name "triad" and was suitable for many prosthodontic applications, including removable, fixed and maxilla facial prostheses. Light-activated urethane dimethacrylate (UDMA) denture base polymers were developed to overcome the allergies, the pungent smell in the laboratory and the conventional time-consuming lost wax method of fabricating prosthesis Polymethyl methacrylate (PMMA) materials [12]. The VLC resin was promoted as a material containing a urethane dimethacrylate matrix with an acrylic copolymer, micro-tins, silica filler, and a photo-initiator system. Triad, as a brand. was the first system of VLC-UDMA denture base polymer. It was presented in the market as a biocompatible material which has ease of fabrication and manipulation, low bacterial adherence, and the ability to bond to other denture base resins [13–16].

Overall, in the literature, the various advantages documented for VLC acrylic resin include rapid preparation time, good rigidity and strength, uniform thickness, and ease of use. Conversely, the disadvantages mentioned are the requirement of a curing unit, which also increases expenditure, stickiness on the surface after curing, difficulty in trimming and poor finish [15–18]. Nevertheless, the main problem reported for VLC record bases is dimensional instability, which is not only affected by material properties and technique, but also by the anatomy of the structure over which the prosthesis is to be made.

One of the major factors affecting dimensional stability is processing shrinkage. The processing shrinkage of the record base is affected by the shape of the palatal concavity and thus shrinkage occurs toward the residual ridge, leading to the lifting of the record base in the mid-palatal region which was cited by Won-suck et al. [8]. The space beneath the posterior palatal region results from processing shrinkage. This dimensional change, that occurs during the denture processing, was recognized by several studies previously and the most possible explanation for it is that the strain release in the maxillary denture base tends to draw the flange inward and the resulting premature contact (gap) of the denture base with the mold in these regions causes the palate to be elevated. Previous research into VLC documented that the material close to the light cure unit would get polymerized first, resulting in gap formation due to transformation and contraction [19]. The cause for the gap formation with the master cast is the contraction because of polymerization and cooling, as well as being lifted by internal stress during polymerization [2,8,12,18]. VLC resins are considered to be dimensionally stable directly after complete polymerization with the use of a proper light source and exposure time [1,2,17,18].

Studies on light-curing resin demonstrated that the nearest material will polymerize first when exposed to polymerizing light, leading to the establishment of a gap due to transformation and contraction [16]. The contraction brought on by polymerization and cooling, as well as the lifting caused by internal stress during polymerization, are the causes of this gap's creation with the master cast [16–18].

Consequently, recommendations have been made in previous studies for improving the accuracy of the VLC resin, whether by a multiple-step curing cycle or by dividing the VLC sheet into two parts along the junction of horizontal and vertical parts of the palate [16,20,21]. By these methods, the stresses that develop at the time of polymerization can be restricted to that segment of the material and shift the direction of shrinkage from the ridge crest area to two separate areas in the middle of the palate and the crest of the ridge [21–23].

Nevertheless, still discrepancies have been reported in the adaptation of visible light cure record bases, despite using different techniques. We decided to explore this area more and the goal set was to formulate an accurate method which can be used on a regular basis with ease to produce VLC record bases with better adaptation so as to accompanied successful prosthesis processes. Thus, this study was conducted with the aim to assess and compare the dimensional stability of light-activated urethane dimethacrylate (UDMA) visible light cure (VLC) denture base, fabricated by different techniques and different curing cycles. The objective was to decide and recommend a technique which gives the most stable record base for regular laboratory use and to determine whether time has any effect on the dimensional stability of VLC record bases.

2. Materials and Methods

The present cross-sectional study was conducted in the Department of Prosthodontics, Career Post Graduate Institute of Dental Sciences and Hospital, Lucknow, India. It was approved by the institute's ethical committee, and the ethical waiver was obtained as there was no involvement of human subjects or tissues (SRC/CPGIDSH/ETHW–11–2019). This study was conducted with the aim of evaluating and comparing the dimensional stability of VLC denture bases fabricated by different techniques and different curing cycles. The objective was to decide on and recommend a technique which gives the most stable record base for regular laboratory use and to determine whether time has any effect on the dimensional stability of VLC record bases.

2.1. Sample Size Calculation

The sample size was selected based on a similar previous study in which the sample size was 30 [12]. Based on the recommendations for sample size calculations in in vitro studies, the number of samples should be in accordance with the statistical calculation of previous studies and should be 10–15% greater than the regular sample size in order to anticipate the incontinence in standards [13]. Thus, in the present study total sample size taken was 45, which was then divided into 3 groups of 15 samples each.

2.2. Fabrication of Samples

A completely edentulous patient was recorded in the outpatient department for denture fabrication. The oral examination was performed for a well-formed, rounded, completely healed edentulous residual alveolar ridge. Patient written consent was obtained before starting the study. The master cast was made after a secondary impression, following the conventional complete denture fabrication protocol.

The master cast obtained in this way was scanned using a desktop laboratory scanner (D800, 3Shape A/S., Copenhagen, Denmark). The laboratory scanner software (2022.1; 3Shape A/S., Copenhagen, Denmark) has a self-alignment setting which joins the multiple scans of various sections and produces a complete virtual image. The data of the scan was stored in a standard tessellation language (STL) format for further use (Figure 1).

The stl file of the master cast was used in an advanced CAD-CAM software [(3-Shape Dental System–Complete Restorative–Complete Denture Module)-2022.1; 3Shape A/S, Copenhagen K Denmark] and printed in castable resin. With the help of a castable die, the metal master die was fabricated in metal [cobalt-chromium (Wirobond C; Bego Gmbh,

Bremen, Germany)] (Figure 1). The metal master die so prepared was duplicated and study casts were poured with type IV die stone (GC Fujirock EP, Kortrijk, Belgium).

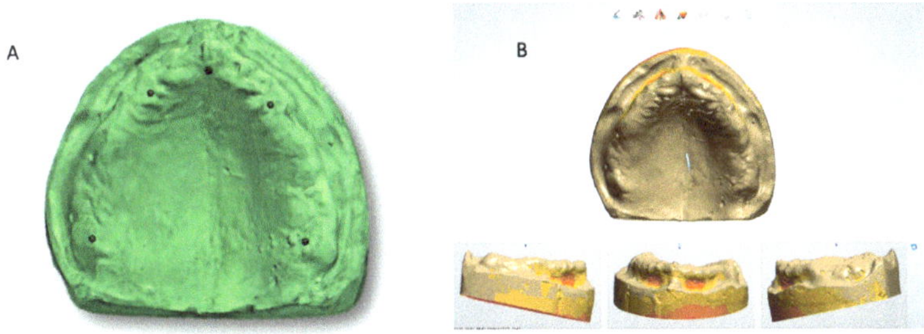

Figure 1. (**A**) Master cast for fabrication of metal master model; (**B**) Representative scanned images of master cast.

A total of 45 casts were made. For each cast, the VLC denture base was fabricated using 3 different techniques. Based on the technique used for the fabrication of denture bases, the samples were divided into 3 groups.

Group A comprised 15 samples which used a conventional technique of fabrication with a short curing cycle of 5 min.

Group B comprised 15 samples which used a new technique of fabrication with a two-step curing cycle (4 + 4 min).

Group C comprised 15 samples which used a two-stage technique of fabrication with multiple steps in the curing cycle (4 + 4 + 2 min).

All the samples were cut at a specific region and observed under a stereomicroscope. The gap/discrepancy between the master cast and VLC record bases at 3 different points was measured. Two times measurement was done after 12 h and 24 h after fabrication.

2.3. Making of Visible Light Cure Record Bases

The VLC record bases were fabricated over the maxillary casts. The VLC sheets were removed from the light protective packet and immediately adapted to the cast by hand pressure. A particular method was used for the adaptation of the VLC sheets. The adaptation started from the centre of the cast toward the periphery in order to reduce air bubbles.

Three techniques were used for making VLC record bases:

(A) Conventional technique—Visible light cure sheets were adapted on the casts and excess material was cut off. Following this, the cast was placed along with the adapted VLC sheet in a light cure polymerizing unit (Tray-Lux) for 5 min (Figure 2).

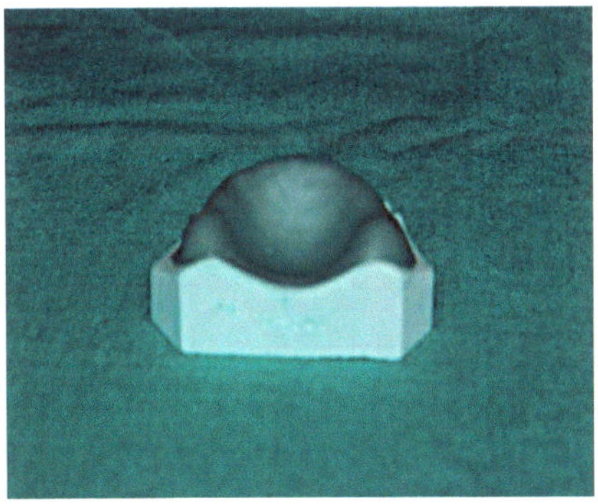

Figure 2. Record Base Fabricated by Conventional Technique (Group-A).

(B) Technique-1—Initially visible light cure sheets were adapted on the cast. Then, a U-shape segment of the adapted VLC sheet was cut along the residual ridge 15 mm below the alveolar crest. The remaining VLC sheet on the cast was then placed in the polymerizing unit for 4 min. The cast was removed from the unit and the cut U-shape segment was replaced on the cast and cured for additional 4 min (Figure 3).

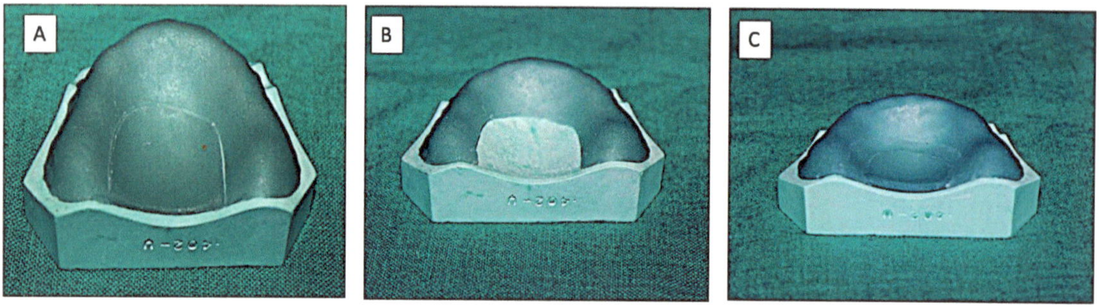

Figure 3. Record Base Fabricated by Technique-1 (Group-B). (**A**) The cut along the residual ridge 15 mm below the alveolar crest; (**B**) The remaining VLC sheet on the cast-cured for 4 min; (**C**) The cut U-shape segment replaced on the cast and cured for additional 4 min.

(C) Technique-2 (two-stage technique)—Visible light cure sheets were adapted on the cast. The excess material was cut from the border. Then, approximately 1.0 mm width of the VLC sheet material was cut at the junction of the horizontal and vertical plane of the palate. The cut was made below 15 mm from the alveolar ridge crest. This was done to obtain two separate components: a U-shaped segment along the residual ridge and an elongated segment in the mid-palatal region (Figure 4). The record base and cast assembly were placed in a polymerizing unit to cure for 4 min. After removing the assembly from the unit, the space between the two segments of the polymerizing sheet was filled using the cut-out VLC sheet. Then, the whole assembly was returned to the polymerizing unit for an additional 4 min. After removing from the unit, an air barrier coating was applied on the internal surface of the record base, and the whole assembly was returned to the polymerizing unit for an additional 2 min.

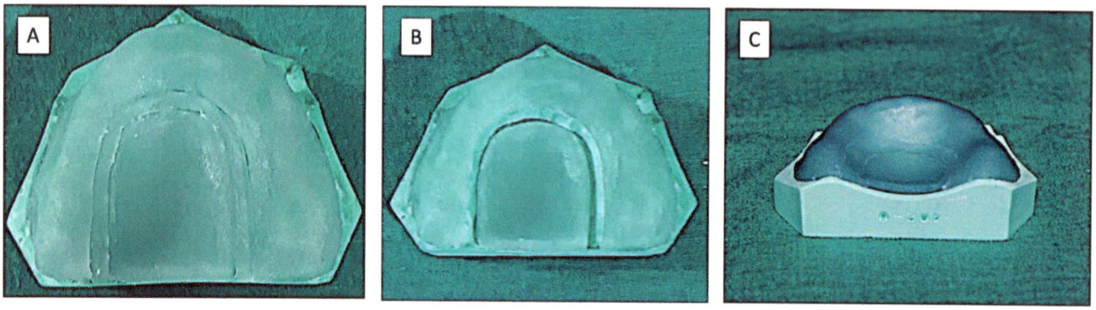

Figure 4. Record Base Fabricated by Technique-2 (Group-C). (**A**) The cut- below 15 mm from the alveolar ridge crest; (**B**) Fist components: a U-shaped segment along the residual ridge and (**C**) second component an elongated segment in the mid-palatal region.

2.4. Sectioning of the Samples

The samples so prepared (cast and VLC record base assembly) were transversely sectioned with the help of a precision cutting machine (BUEHLER Worldwide, Waukegan Road Lake Bluff, IL, USA). The cuts were made at predetermined marked locations (canine region, molar region, posterior palatal seal region). In each region, 3 points (1- right ridge crest; 2- left ridge crest and 3- mid-palate region) were marked to measure the discrepancy:

The canine region (c), [right ridge crest (c1), left ridge crest (c2) and mid-palate region (c3)]; The molar region (m), [right ridge crest (m1), left ridge crest (m2), and mid-palatal region (m3)]

The posterior palatal seal region (p) [right ridge crest (p1), left ridge crest (p2), and posterior palatal seal region (p3)] (Figure 5).

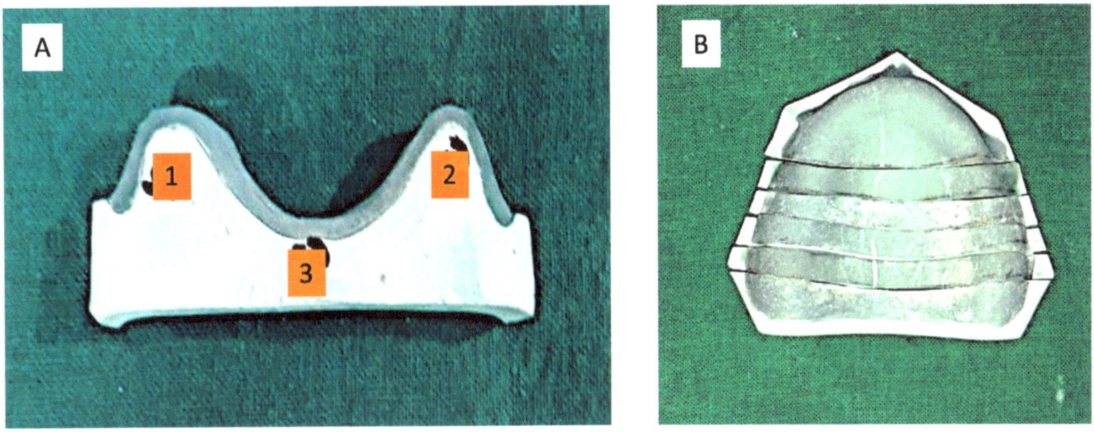

Figure 5. Sectioning & Marking of the Cast & Record Base Assembly (**A**) Coronal section view (1—right ridge crest; 2—left ridge crest and 3—mid-palate region); (**B**) Occlusal view.

2.5. Measuring the Discrepancy between Cast and Record Bases by Using a Stereomicroscope

After cutting all the samples, the discrepancy was measured at 9 pre-marked locations. The first measurement was performed after 12 h and the second after 24 h of record base fabrication. For evaluating the discrepancy, the perpendicular distances were measured from the internal surface of the record base to the external surface of the cast. Samples were studied under a stereomicroscope under a magnification of 20×, which can measure up to 0.001 mm (Figure 6). At each point, the discrepancy reading was recorded twice, and the average was taken as the final reading. All measurements were recorded by a trained technician who was blinded from the study. All measurements were recorded and analyzed using statistical analysis software (IBM SPSS Statistics for Windows v22; IBM Corp, Armonk, NY, USA).

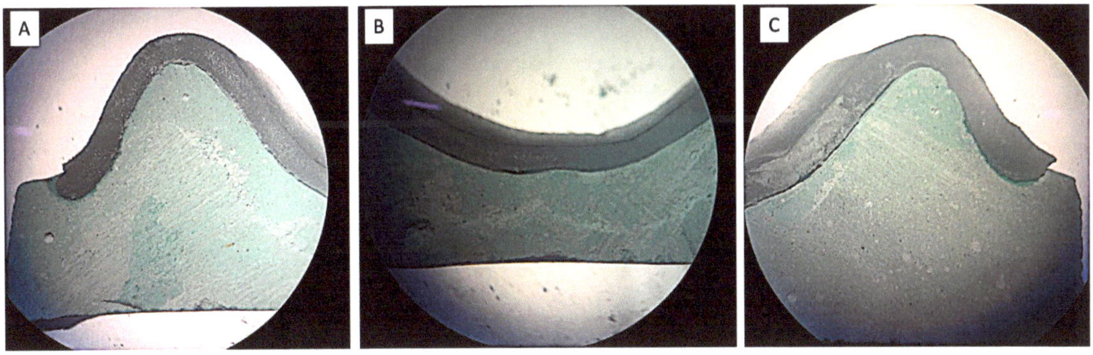

Figure 6. Discrepancy Seen Under the Stereomicroscope at (**A**) Right Ridge Crest; (**B**) Mid-Palate Region; (**C**) Left Ridge Crest.

2.6. Statistical Analysis

Microsoft Excel was used to calculate the means, standard deviation, and standard error of the mean (SEM). Two-way ANOVA, Tukey HSD and a t test was conducted to measure the statistical significance of differences among the groups. The Statistical Package for the Social Sciences (SPSS) software version 22 IBM.; Chicago (Ill., USA) was used to find the significant difference between the techniques. The significance level was set to 95%.

3. Results

The results of the study showed that there was a discrepancy between the record base and cast. The results were studied region- and time-wise.

3.1. Dimensional Discrepancy between Record Base and Cast after 12 h

The dimensional discrepancy of three groups (Group A, Group B and Group C) at three different sites (right ridge crest, left ridge crest and mid-palate) of the canine, region after 12 h was significantly different among the groups at both right ridge crest (F = 6.69, $p = 0.003$) and mid-palate (F = 10.34, $p < 0.001$), whereas the difference was insignificant at left ridge crest (F = 0.68, $p = 0.512$). For the molar region, there was a significantly different dimensional discrepancy among the groups at all three sites; right ridge crest (F = 40.01, $p < 0.001$), left ridge crest (F = 34.69, $p < 0.001$) and mid-palate (F = 67.81, $p < 0.001$). For the posterior palatal seal, the region showed significantly different dimensional discrepancies among the groups at all three sites; right ridge crest (F = 23.38, $p < 0.001$), left ridge crest (F = 23.27, $p < 0.001$) and mid-palate (F = 67.36, $p < 0.001$) (Table 1).

Table 1. For each site of canine region, comparisons of difference in mean dimensional discrepancy (mm) between groups by Tukey test at 12 h.

Anatomical Regions	Comparison	Right Ridge Crest		Left Ridge Crest		Mid-Palate	
		Mean Diff.	p Value	Mean Diff.	p Value	Mean Diff.	p Value
Canine region	Group A vs. Group B	0.062	0.012	0.009	0.789	0.099	0.001
	Group A vs. Group C	0.068	0.006	0.015	0.481	0.096	0.001
	Group B vs. Group C	0.006	0.954	0.007	0.869	0.003	0.990
Molar region	Group A vs. Group B	0.099	0.003	0.090	0.002	0.191	<0.001
	Group A vs. Group C	0.253	<0.001	0.208	<0.001	0.454	<0.001
	Group B vs. Group C	0.154	<0.001	0.118	<0.001	0.263	<0.001
Posterior palatal region	Group A vs. Group B	0.033	0.745	0.061	0.365	0.090	0.211
	Group A vs. Group C	0.284	<0.001	0.289	<0.001	0.566	<0.001
	Group B vs. Group C	0.251	<0.001	0.228	<0.001	0.476	<0.001

Further, for each site, comparing the difference in the mean dimensional discrepancy between the groups, the Tukey test showed a significantly ($p < 0.05$ or $p < 0.01$) different and lower dimensional discrepancy in both Group B and Group C as compared to Group A at all sites for canine and molar region. Furthermore, at all sites, it was also found to be significantly ($p < 0.001$) different and lower in Group C compared to Group B (Figure 7).

Tukey test showed significantly ($p < 0.001$) different and lower dimensional discrepancy in Group C, as compared to both Group A and Group B, at all sites. However, at all sites, it did not differ ($p > 0.05$) between Group A and Group B, i.e., found to be statistically the same (Figure 7).

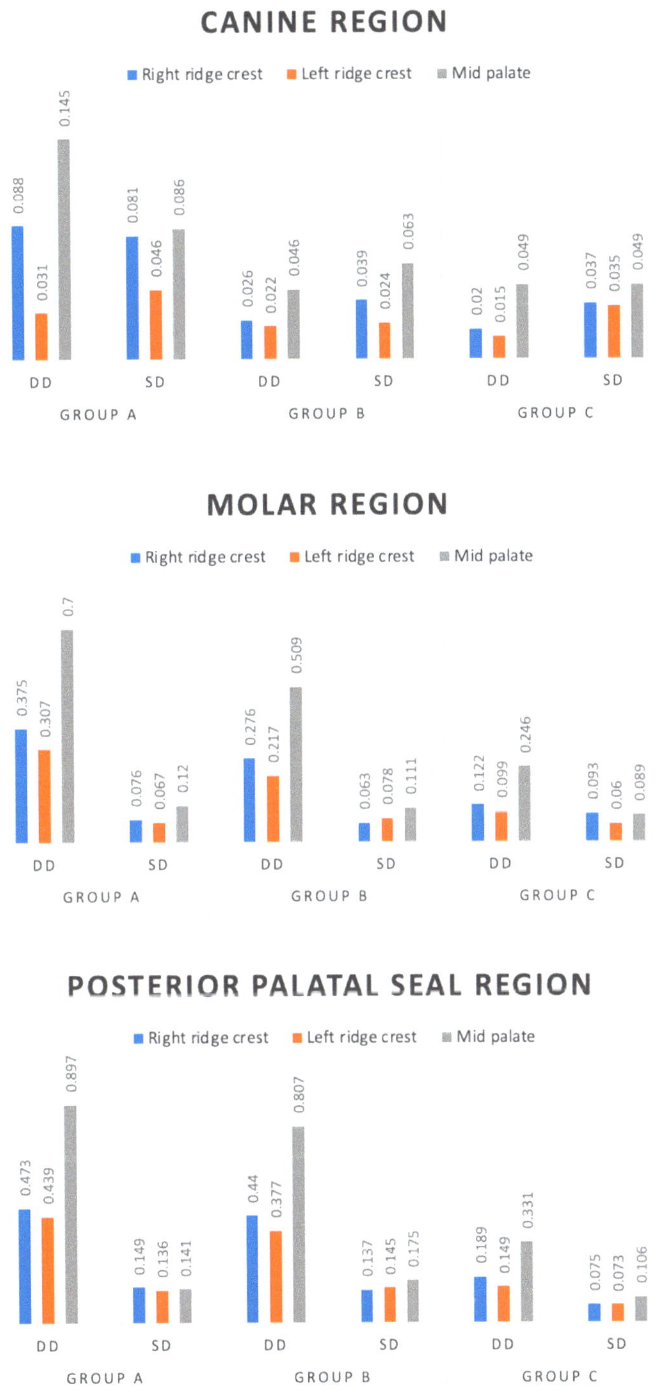

Figure 7. Mean dimensional discrepancy (mm) of three groups at three sites of canine, Molar and Posterior Palatal region at 12 h. Dimensional discrepancy (DD); Standard deviation (SD).

3.2. Dimensional Discrepancy between Record Base and Cast after 24 h

The dimensional discrepancy of three groups (Group A, Group B and Group C) at three different sites (right ridge crest, left ridge crest and mid-palate) of the canine region after 24 h showed significantly different dimensional discrepancies among the groups at all three sites; right ridge crest (F = 71.42, $p < 0.001$), left ridge crest (F = 81.84, $p < 0.001$) and mid-palate (F = 59.21, $p < 0.001$). For the molar region, significantly different dimensional discrepancies among the groups at all three sites; right ridge crest (F = 93.54, $p < 0.001$), left ridge crest (F = 105.96, $p < 0.001$) and mid-palate (F = 125.53, $p < 0.001$). For the posterior palatal seal region, significantly different dimensional discrepancies among the groups at all three sites; right ridge crest (F = 96.97, $p < 0.001$), left ridge crest (F = 102.24, $p < 0.001$) and mid-palate (F = 14.56, $p < 0.001$).

The Tukey test showed significantly ($p < 0.05$ or $p < 0.01$ or $p < 0.001$) different and lower dimensional discrepancies in both Group B and Group C as compared to Group A at all sites in both canine and molar regions (except Group A and Group B in mid-palate in the molar region.). Furthermore, at all sites, it also lowered significantly ($p < 0.001$) in Group C as compared to Group B. Moreover, in posterior palatal seal region, at both the right and left ridge crest, dimensional discrepancy also got lowered significantly ($p < 0.001$) in Group C as compared to Group B but at mid-palate, it did not differed ($p > 0.05$) between the two groups, i.e., found to be statistically the same (Table 2, Figure 8).

Table 2. For each site of the canine region, comparisons of difference in mean dimensional discrepancy (mm) between groups by Tukey test at 24 h.

Anatomical Regions	Comparison	Right Ridge Crest		Left Ridge Crest		Mid-Palate	
		Mean Diff.	*p* Value	Mean Diff.	*p* Value	Mean Diff.	*p* Value
Canine region	Group A vs. Group B	0.075	<0.001	0.051	0.002	0.060	0.042
	Group A vs. Group C	0.206	<0.001	0.174	<0.001	0.249	<0.001
	Group B vs. Group C	0.131	<0.001	0.123	<0.001	0.189	<0.001
Molar region	Group A vs. Group B	0.159	<0.001	0.145	<0.001	0.049	0.210
	Group A vs. Group C	0.381	<0.001	0.351	<0.001	0.411	<0.001
	Group B vs. Group C	0.223	<0.001	0.206	<0.001	0.362	<0.001
Posterior palatal region	Group A vs. Group B	0.133	<0.001	0.130	<0.001	0.805	0.003
	Group A vs. Group C	0.427	<0.001	0.381	<0.001	1.227	<0.001
	Group B vs. Group C	0.294	<0.001	0.251	<0.001	0.422	0.173

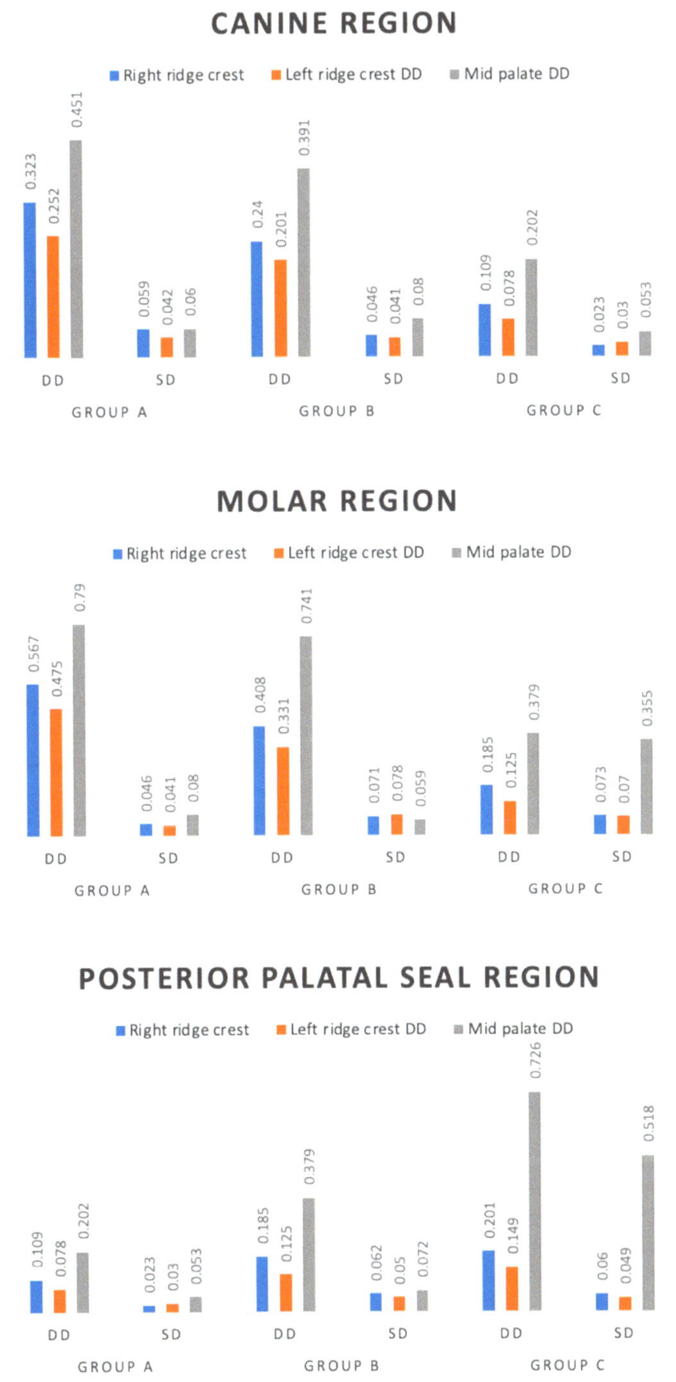

Figure 8. Mean dimensional discrepancy (mm) of three groups at three sites of canine, Molar and Posterior Palatal region at 24 h. Dimensional discrepancy (DD); Standard deviation (SD).

4. Discussion

Acrylic resins are widely used in dentistry to fabricate custom trays and record bases. These are used for making final impressions and recording maxillo-mandibular relations and arranging artificial teeth. To ensure intra-oral stability and the retention of the record base, the base should be closely adapted to the cast and should be dimensionally stable [18].

The introduction of auto-polymerizing acrylate polymers brought significant improvement to the mechanical properties of custom trays and record bases, but the use of self-cure resin has certain disadvantages such as shrinkage during polymerization, toxicity, the release of pungent vapors, residual monomer and adverse tissue reactions and related diseases. The improvement in the field of material science, especially in the field of dental materials, and the development of newer forms of denture base materials resulted in better denture base resins and helped in overcoming some of these drawbacks.

For example, the irritation to the mucosa and polymerization shrinkage is overwhelmed by polycarbonate, nylon and light-activated urethane dimethacrylate (UDMA) denture-base polymer. This VLC material is very common nowadays and is widely accepted both by patients and dentists because of its convenience to use, lack of monomers and superior dimensional stability after fabrication [24]. Additionally, its hazardous effects are very limited, the laboratory time required is significantly reduced time, and it has ease of use, good accuracy, strength and rigidity, uniform thickness, and good dimensional stability [25]. Visible light-cured resin is characterized by improved physical characteristics, such as increased stiffness, good form, volume stability, and low sensitivity to moisture. In addition, these materials are easy to use, and their use saves time [10–12]. Hence, the record base should be fabricated to be accurately fitted, and it should not change, but remain stable. Even the inaccurate record bases can look well fitted in the palate because of the elasticity of soft tissue, but it does not mean that it fits accurately in reality. The inaccurate record base incurs errors in the final occlusion of the complete denture due to the inaccurate jaw relation record [12,14]. The formation of the gap during the adaptation of the record base is very common in the posterior border of the maxillary cast. More specifically, it is formed in the posterior palatal seal area [21].

Different techniques have been suggested by various authors to overcome these defects and to obtain a more accurate fit. These techniques include the record base with elastic impression materials, vacuum adaptation of the record base, and staged polymerization by covering the palatal portion of the record base. However, the problem with resilient liners is tearing/detachment from the record base surface due to repeated insertion and removal, causing the inaccurate seating of the bases. The use of vacuum only improves adaptation, without ever addressing gap formation resulting from polymerization shrinkage [2].

The shrinkage of denture base acrylic resin is acknowledged as an error in the material; it results in space existing beneath the posterior palatal region. The most likely explanation for this dimension change, which has been noted in previous studies, is that the strain release in the maxillary denture base tends to draw the flange inward. As a result, there is an early contact (gap) of the denture base with the mold in these areas, which elevates the palate. These findings also show that acrylic resin's inherent lack of dimensional precision is one of the potential causes of denture discrepancies [22,23,26–28].

Previous studies conducted comparisons of the self-cure and heat cure resins for adaptation using various techniques and determined that the pour technique is still a viable alternative for clinicians and dental technicians, as the fit of the denture is similar to that of the compression-moulding and injection-moulding techniques [19]. Even after three weeks of water storage, the residual monomer content of the heat-curing polymers is reduced but otherwise similar to that of the cold polymers [20]. These denture base acrylic resins will always contain an unreacted residual monomer that could irritate people and trigger allergies, regardless of the curing method used [22]. Heat-curing polymers also exhibit greater fracture resistance and less polymerization shrinkage, but they also exhibit higher stresses during the subsequent cooling phase.

The present study was conducted to evaluate and compare the dimensional discrepancy between cast and record bases using visible light cure material by different techniques and curing cycles at different time intervals. The results of the study showed that dimensional accuracy or fit of the denture base processed by technique 2, the two-stage technique with a multiple-step curing cycle, was better than that processed by the conventional curing cycle, and technique 1. Thus, the null hypothesis formulated was rejected. In the present study, the master metal model was made by using a castable model generated by 3-D printing, which helped in making uniform sample casts, thus eliminating the chance of error at the cast level.

Record bases, fabricated by a conventional method with a single-step curing cycle of 4 min showed more dimensional discrepancy due to the shape of the palatal concavity. Shrinkage occurred toward the residual ridge, leading to the lifting of the record base in the mid-palatal region, an occurrence which was cited by Won-suck et al. [2].

In techniques 1 and 2, i.e., with two steps curing cycle (4 min + 4 min) and a two-stage technique with a multiple-step curing cycle (4 min + 4 min + 2 min), the VLC sheet was cut along the residual ridge. Thus, the accuracy of the light-curing resin can be improved through multiple-step curing methods or by dividing the VLC sheet into two pieces along the junction of the horizontal and vertical configurations of the palate [21]. This method of cutting the VLC sheet into pieces and staged curing produced less discrepancy, which is a result of dividing the VLC resin sheets into two pieces along the palatal portion at a constant distance from the alveolar crest. By doing so, the distribution of contraction forces was performed evenly and applied stress to each of the pieces.

The edentulous maxilla consists of a relatively flat portion in the middle of the hard palate and inclined slopes towards the residual ridge. Due to the shape of palatal concavity, shrinkage occurs toward the residual ridge leading to the lifting of the record base in the mid-palatal region, as reported by Won-suck et al. [2].

The stresses that form during polymerization can be contained within each segment of the VLC sheet by cutting it into two pieces at the point where the palate's horizontal and vertical configurations meet. This allows for the direction of shrinkage to be changed from the ridge crest area to two distinct areas in the middle part of the palate and at the crest. This study agrees with the study of Kenneth et al., in which it was described that the gap formation due to polymerization shrinkage could be reduced by limiting the amount of composite exposed to the curing light at one time and readjusting the uncured composite to the cast between curing cycles. Theoretically, compensation for discrepancies might be made for a record base, with a dimensional change of that proportion resting on compressible tissue without jeopardizing stability [21]. To achieve greater adaptability, it is advised that VLC record bases be cured using a two-stage approach with a multiple-step curing cycle based on the findings of this study

In the present study, at the canine region, molar region, and posterior palatal seal region, the dimensional discrepancy was found to be the highest at the mid-palate region for all the groups (A, B, C) followed by the right ridge crest, and it was found to be lowest in left ridge crest. At the canine region among all the groups, a significant dimensional discrepancy was found at both the right ridge crest (c1) and mid-palate region (c3) and it was found to be insignificant at the left ridge crest (c2). Both Group B and Group C have shown a low dimensional discrepancy, as compared to Group A at both the right ridge crest (c1) and mid-palate (c3), which was significant. However, at the left ridge crest (c2) it was found to be similar between the groups after 12 h. The dimensional discrepancy was found to be similar at all sites (c1, c2, c3) between Group B and Group C after 12 h. These values are based on a previous study which is done by Kenneth G. Boberick et al. [21].

At the canine region, dimensional discrepancy was found significantly different ($p < 0.001$) at all three sites (c1, c2, c3) among all the groups after 24 h. In both Group B and Group C, the low and significant dimensional discrepancy was found at all sites as compared to Group A and the least dimensional discrepancy was found in Group C at all sites (c1, c2, c3) after 24 h.

At the molar region for all sites (m1, m2, m3), among all groups, a significant dimensional discrepancy was found ($p < 0.001$) after 12 h. A mid-palate region (m3) showed the highest discrepancy as compared to right ridge crests (m1) and left ridge crest (m2) among all groups. Due to the shape of the palatal concavity, shrinkage occurs toward the residual ridge, leading to the lifting of the record base in the mid-palatal region cited by Won-suck et al. [2].

Consani et al. suggested that the molar region is the most important site for gap space production between the palatal zone and the record base due to linear shrinkage [10]. At the molar region, in both Group B and Group C, the low dimensional discrepancy was found as compared to group A at all sites which were found significant ($p < 0.001$). This may be due to dividing the VLC sheet into two pieces along the junction of the horizontal and vertical configuration of the palate. The stresses that develop during polymerization can be confined within each segment of the material and lead to the direction of shrinkage shifting from the ridge crest area to two separate areas in the middle of the palate and the crest of the ridge [10].

At all sites, Group C was found to have the least dimensional discrepancy at the molar region as compared to Group A, and Group B was found to be significant after 12 h. This may be because of the light-polymerizing resin being divided into more pieces along the palatal portion by distributing force and limiting the amount of surface area exposed to the curing light and readapting the uncured visible light cure sheet to the cast between curing episodes [12]. At all the sites, in both Group B and Group C, mean dimensional discrepancy was found low and significant as compared to Group A after 24 h and it was found to be increased after 24 h as compared to the 12 h mark in the molar region.

At the posterior palatal seal region, for all sites (p1, p2, p3) dimensional discrepancy was found to be lower and significant in Group C as compared to Group A and Group B after 12 h. However, between Group A and Group B, it was found statistically insignificant after 12 h. For all sites (p1, p2, p3), dimensional discrepancy was found to be less significant in Group B and Group C as compared to Group A after 24 h and it was found to be more significant as compared to 12 h at all sites. In Group C at the mid-palate point, it was found statistically similar as compared to Group B after 24 h.

At the posterior palatal seal region, the dimensional discrepancy was found highest at all sites among all the regions. The greatest discrepancy occurred at the mid-palatal region of the posterior border, in the area of the posterior palatal seal, and gradually decreased anteriorly in agreement with the previous study [21]. The results of this study revealed a discrepancy pattern between the record base and cast, which is not uniform over the palatal surface [21].

In the present study, Group A showed the highest discrepancy as compared to Group B and Group C, at all the sites for all the regions Group B and Group C have shown low dimensional discrepancy of all the regions at a different site and it is found to be lowest with Group C after 12 and 24 h.

Therefore, it is recommended that the accuracy of the light-curing resin should be improved through multiple-step curing methods or by dividing the VLC sheet into two pieces along the junction of horizontal and vertical configurations of the palate. The completeness of polymerization is also a significant factor for two reasons: first, the degree of polymerization affects the geometric and material properties of the resultant prosthesis [29]. Secondly, the unreacted monomer may produce undesirable effects on the human body [30]. However, it has been reported in the literature that polymerized UDMA denture bases are nontoxic and that unpolymerized VLC materials have very low toxicity [31]. UDMA material has been proven to be less allergenic than other acrylics [32–34]. The recommendations for VLC denture bases by various authors are not only based on satisfactory strength and dimensional stability but also on ease of manipulation and fabrication [31–36].

Even though the study was conducted in a precise way, there are certain limitations to its usability, which include the limited total sample size. The cutting of cast and incorporation of dust created problems while determining the discrepancies. It is recommended to use a new technique of micro-computer tomography for the evaluation of the gap below the record base without cutting the samples. The effect of light absorption from the cast, which was green in colour, was not taken into consideration as it was reported in a previous study [16] that the dimensional accuracy of the VLC trial denture base could be improved by using a dark-coloured cast. Additionally, comparisons with the newer advanced CAD-CAM milled and 3D printed material was not done, which is recommended for future studies.

5. Conclusions

Within the limitations of the study, the following conclusion can be drawn: the most accurate light-activated urethane dimethacrylate (UDMA) visible light cure (VLC) denture base can be obtained by a two-stage technique with multiple steps curing cycles (as in Group C). The dimensional discrepancy increases with time, irrespective of the techniques used for base fabrication (record bases were found to be dimensionally unstable after 24 h).

Author Contributions: Conceptualization, S.M., S.C., M.A. and K.K.P.; methodology, S.M., S.C., M.A., K.K.P., N.M.A. and M.A.A.; software, S.M., S.C., M.A., K.K.P., N.M.A., M.A.A., M.K.A., S.K.V., N.M.A.A., S.M.A., A.Y. and W.M.S.A.; validation, S.M., S.C., M.A., K.K.P., N.M.A., M.A.A., M.K.A., S.K.V., N.M.A.A., S.M.A., A.Y. and W.M.S.A.; formal analysis, S.M., S.C., M.A. and K.K.P.; investigation, S.M., S.C., MA. and K.K.P.; resources, S.M., S.C., M.A., K.K.P., N.M.A., M.A.A., M.K.A., S.K.V., N.M.A.A., S.M.A., A.Y. and W.M.S.A.; data curation, S.M., S.C., M.A. and K.K.P.; writing—original draft preparation, S.M., S.C., M.A., K.K.P., N.M.A., M.A.A., M.K.A., S.K.V., N.M.A.A., S.M.A., A.Y. and W.M.S.A.; writing—review and editing, S.M., S.C., M.A., K.K.P., N.M.A., M.A.A., M.K.A., S.K.V., N.M.A.A., S.M.A., A.Y. and W.M.S.A.; visualization, S.M., S.C., M.A., K.K.P., N.M.A., M.A.A., M.K.A., S.K.V., N.M.A.A., S.M.A., A.Y. and W.M.S.A.; supervision, S.M., S.C., M.A., K.K.P., N.M.A., M.A.A., M.K.A., S.K.V., N.M.A.A., S.M.A., A.Y. and W.M.S.A.; project administration, S.M., S.C., M.A., K.K.P., N.M.A. and M.A.A.; funding acquisition, S.M., S.C., M.A., K.K.P., N.M.A., M.A.A., M.K.A., S.K.V., N.M.A.A., S.M.A., A.Y. and W.M.S.A. All authors have read and agreed to the published version of the manuscript.

Funding: The authors extend their appreciation to the Deanship of Scientific research at King Khalid University, Saudi Arabia for funding this work through Small Group Project under the grant number (RGP.1/204/43).

Institutional Review Board Statement: Ethical waiver was obtained in this study as no human tissues were involved (SRC/CPGIDSH/ETHW–11–2019).

Data Availability Statement: Data can be made available on demand by the chief researcher for academic purposes by email.

Conflicts of Interest: The authors declare no conflict of interest. The funders had no role in the design of the study; in the collection, analyses, or interpretation of data; in the writing of the manuscript, or in the decision to publish the results.

References

1. Zissis, A.; Huggett, R.; Harrison, A. Measurement method used for the determination of dimensional accuracy and stability of denture base material. *J. Dent.* **1991**, *19*, 199–206. [CrossRef] [PubMed]
2. Oh, W.-S.; May, K.B. Two stage technique for optimum fit and stability of light-polymerized record bases. *J. Prosthet. Dent.* **2008**, *99*, 410–411. [CrossRef] [PubMed]
3. Chaturvedi, S.; Addas, M.K.; Alqahtani, N.M.; Al Ahmari, N.M.; Alfarsi, M.A. Computerized occlusal forces analysis in complete dentures fabricated by additive and subtractive techniques. *Technol. Health Care* **2021**, *29*, 781–795. [CrossRef]
4. Bonatti, M.R.; Cunha, T.R.; Regis, R.R.; Silva-Lovato, C.H.; Paranhos, H.D.; de Souza, R.F. The effect of polymerization cycles on color stability of microwave-processed denture base resin. *J. Prosthodont.* **2009**, *18*, 432–437. [CrossRef] [PubMed]
5. Dimitrova, M.; Corsalini, M.; Kazakova, R.; Vlahova, A.; Chuchulska, B.; Barile, G.; Capodiferro, S.; Kazakov, S. Comparison between Conventional PMMA and 3D Printed Resins for Denture Bases: A Narrative Review. *J. Compos. Sci.* **2022**, *6*, 87. [CrossRef]

6. Wong, D.; Cheng, L.Y.; Chow, T.W.; Clark, R.K. Effect of processing method on the dimensional accuracy and water sorption of acrylic resin denture. *J. Prosthet. Dent.* **1999**, *81*, 300–304. [CrossRef] [PubMed]
7. Chaturvedi, S.; Addas, M.K.; Alqahtani, N.M.; Al Ahmari, N.M.; Alfarsi, M.A. Clinical analysis of CAD-CAM milled and printed complete dentures using computerized occlusal force analyser. *Technol. Health Care* **2021**, *29*, 797–811. [CrossRef]
8. Bartoloni, J.A.; Murchison, D.F.; Wofford, D.T.; Sarkar, N.N. Degree of conversion in the denture base materials for varied polymerization techniques. *J. Oral Rehabil.* **2000**, *27*, 488–493. [CrossRef]
9. Crag, R.G.; Powers, G.M. *Restoration Dental Material*, 11th ed.; The Mosbey: St. Louis, MO, USA, 2002; pp. 500–540.
10. Consani, R.L.X.; Domitti, S.S.; Rizzatti Barbosa, C.M.; Consani, S. Effect of commercial acrylic resin on dimensional accuracy of the maxillary denture base. *Braz. Dent. J.* **2002**, *13*, 57–60.
11. Consani, R.L.X.; Paula, A.B.; Fugolin, A.P.P.; Pfeifer, C.S. Strategies for Potential Toughening of Acrylic Denture Bases Polymerized with Microwave Energy. *Braz. Dent. J.* **2020**, *31*, 523–531. [CrossRef]
12. Fatihallah, A.A.; Mohamed, M.R.; Zia, K.G. Evaluation of dimensional stability for denture bases in different curing techniques. *Mustansiria Dent. J.* **2009**, *6*, 389–394. [CrossRef]
13. Chander, N.G. Standardization of in vitro studies. *J. Indian Prosthodont. Soc.* **2016**, *16*, 227–228. [PubMed]
14. Diaz-Arnold, A.M.; Vargas, M.A.; Shaull, K.L.; Laffoon, J.E.; Qian, F. Flexural and fatigue strengths of denture base resin. *J. Prosthet. Dent.* **2008**, *100*, 47–51. [CrossRef]
15. Elkholy, S. Improving Dimensional Stability of Visible Light Cure Trail Denture Base using Different Colored Cast. *J. Contemp. Dent.* **2016**, *6*, 171–176. [CrossRef]
16. Huh, B.J.; Kang, G.M.; Shin, W.S.; Ryu, J.J. Effect of two-phase fabrication method for the optimum fit of light-polymerized record bases. *J. Adv. Prosthodont.* **2010**, *2*, 102–105. [CrossRef] [PubMed]
17. Anulekha, K.; Parameshwari, G.; Taruna, M.; Tella, S.; Reddy, R.; Jagani, A.S. A Comparative Evaluation of the Denture Base Accuracy along the Posterior Palatal Seal using Different Anchorage Systems in Complete Dentures. *J. Adv. Med. Dent. Sci. Res.* **2016**, *4*, 165–169.
18. Choi, H.J.; Lim, J.H.; Jo, I.H. A study on the accuracy of the record base of the complete denture to the master cast according to kinds of resin and polymerization method. *J. Korean Acad. Prosthodont.* **2000**, *38*, 310–320.
19. Wirz, J.; Jaeger, K.; Schmidli, F. Light-polymerized materials for custom impression trays. *Int. J. Prosthodont.* **1990**, *3*, 64–71.
20. Andreopoulos, A.G.; Polyzois, G.L. Repair of denture base resins using visible light-cured materials. *J. Prosthet. Dent.* **1994**, *72*, 462–468. [CrossRef]
21. Boberick, K.G.; McCool, J. Dimensional stability of record bases fabricated from light-polymerized composite using two methods. *J. Prosthet. Dent.* **1998**, *79*, 399–403. [CrossRef]
22. Faot, F.; da Silva, W.J.; da Rosa, R.S.; Del Bel Cury, A.A.; Garcia, R.C. Strength of denture base resins repaired with auto- and visible light-polymerized materials. *J. Prosthodont.* **2009**, *18*, 496–502. [CrossRef]
23. Pronych, G.J.; Sutow, E.J.; Sykora, O. dimensional stability and dehydration of a thermoplastic polycarbonate based and two PMMA based denture resins. *J. Oral Rehabil.* **2003**, *30*, 1157–1161. [CrossRef] [PubMed]
24. Kazanji, M.; Majeed, A. The effect of storage and curing time on dimensional changes of visible light cured acrylic denture base (VLCADB). *IOSR J. Dent. Med. Sci.* **2015**, *14*, 24–26.
25. Elkholy, S. Evaluation of dimensional stability of visible light cured custom tray for complete denture. *Egypt. Dent. J.* **2011**, *57*, 1585–1593.
26. Takamata, T.; Setcos, J.C.; Philips, R.W.; Boone, M.E. Adaptation of acrylic resin dentures as influenced by the activation mode of polymerization. *J. Am. Dent. Assoc.* **1989**, *119*, 271–276. [CrossRef] [PubMed]
27. Al-Mulla, M.A.S.; Huggett, R.; Brooks, S.C.; Murphy, W.M. Some physical and mechanical properties of a visible light-activated material. *Dent. Mater.* **1988**, *4*, 197–200. [CrossRef] [PubMed]
28. Danesha, G.; Lippolda, C.; Mischke, K. Polymerization characteristics of light- and auto-curing resins for individual splints. *Dent. Mater.* **2006**, *22*, 426–433. [CrossRef] [PubMed]
29. Gorman, C.M.; O'Sullivan, M. Fabrication of a duplicate denture using visible light-polymerized resin as an interim denture base. *J. Prosthet. Dent.* **2006**, *96*, 374–376. [CrossRef] [PubMed]
30. Rode, S.M.; Cavalcanti, B.N.; Ferrielo, V.; Faria, M.R.; Villa, N. Biocompatibility of Two Types of resins for Prosthetic Usage. *Cienc. Odontol. Bras.* **2007**, *10*, 6–9. [CrossRef]
31. Aoyama, Y.; Maeda, M.; Yokoi, R.; Saitoh, A.; Tsukimura, N.; Shimada, A.; Satoh, T.; Hayakawa, J.; Takeda, T.; Ishigami, K.; et al. Basic studies on visible light-curing resin for denture base. Part 8. The hardening depth of denture relining material after penetration of light through the denture base resin. *J. Nihon Univ. Sch. Dent.* **1991**, *33*, 73–80. [CrossRef]
32. Elahi, J.M.; Abdullah, M.A. Effect of different polymerization techniques on dimensional stability of record bases. *J. Prosthet. Dent.* **1994**, *71*, 150–153. [CrossRef] [PubMed]
33. Alkurt, M.; Yeşil Duymuş, Z.; Gundogdu, M. Effect of repair resin type and surface treatment on the repair strength of heat-polymerized denture base resin. *J. Prosthet. Dent.* **2014**, *111*, 71–78. [CrossRef] [PubMed]
34. Khan, S.B.; Geerts, G. Determining the dimensional stability, fracture toughness and flexural strength of light-cured acrylic resin custom tray material. *Eur. J. Prosthodont. Restor. Dent.* **2009**, *17*, 67–72. [PubMed]

35. Chaturvedi, S.; Khaled, A.M.; Al Humaidi, A.S.A.; Al Qahtani, A.M.; Al Qahtani, M.D. A Novel Approach to Determine the Prevalence of Type of Soft Palate Using Digital Intraoral Impression. *Int. J. Dent.* **2017**, *2017*, 3268064. [CrossRef]
36. Chung, K.H.; Wadhwani, C.; Cagna, D.R. A simple technique to control the heat generated during light polymerization of custom impression trays. *J. Prosthet. Dent.* **2014**, *111*, 79–80. [CrossRef]

polymers

MDPI

Article

Emulsion Nanofibres as a Composite for a Textile Touch Sensor

David Mínguez-García, Pablo Díaz-García , Jaime Gisbert-Payá and Marilés Bonet-Aracil *

Departamento de Ingeniería Textil y Papelera, Universitat Politècnica de València, 03801 Alcoy, Spain; damingar@epsa.upv.es (D.M.-G.); pdiazga@txp.upv.es (P.D.-G.); jaigispa@txp.upv.es (J.G.-P.)
* Correspondence: maboar@txp.upv.es; Tel.: + 34-966528470

Abstract: The combination of a nanofibre net and textile support represents an interesting composite capable of conferring various properties. Nanofibres are so thin that they can be easily damaged by human touch. In this study, we hypothesised that dyeing nanofibres with different colours from their textile supports would result in a colour difference upon their degradation, providing evidence that the composite has been touched and acting as a touch sensor. Two different methods were studied: directly inserting the dye into the polymer via electrospinning or creating a coloured liquid emulsion encapsulated by the polymer via electrospinning. Two black dyes were studied. Colour index (CI) Acid Black 194 was added directly to polyvinyl alcohol (PVA) as the polymer. Sage oil was used for CI Solvent Black 3. The nanofibre nets were conveniently electrospun on a white polyester fabric; the fabrics were then characterised by colour coordinate analysis, FTIR, and SEM. The results showed that the dyed solution in oil was encapsulated, and the black colour could only be observed when rubbed, whereas the dyed polymer showed a black colour that was removed when rubbed. Therefore, the hypothesis was confirmed, and both samples demonstrated the desired touch sensor behaviour.

Keywords: electrospun; polyvinyl alcohol; fibre; colour

1. Introduction

The application of nanofibrous veils in multiple industrial sectors has increased dramatically in recent decades. Interest in electrospun nanofibres has arisen due to their exceptional characteristics, high surface-area-to-volume ratio, high porosity, ultra-thin interconnected and lightweight fibrous structure, and their potential for use with a multitude of synthetic and natural polymers and compounds that provide different characteristics [1]. Nanofibre veils have been widely studied in filtration-related fields, including air filtration [2–4], water filtration [5–7], biomedicine [8–10], sensors [11,12], soundproofing [13,14], defence and protection garments [15,16], food packaging [17,18], cosmetics [19,20], and home furnishing [21,22].

Currently, across various fields of application, there is a growing need for the creation of a singular matrix that possesses multiple functional traits. By integrating functional nanoparticles and active ingredients into electrospun fibres, it becomes possible to fulfil the demands for enhanced mechanical strength, electrochemical properties, and other technical performances all at once. In the food sector, authors have combined thymol essential oil with PLGA nanofibres via coaxial electrospinning to prevent microbial growth in fruits during storage [23]. In the biomedicine field, the addition of camptothecin (CPT), a natural hydrophobic chemotherapy agent, to electrospun amphiphilic peptide (AP) nanofibres has demonstrated the inhibition of breast cancer tumour growth [24]. Nanofibre veils have also been used in the design of protective clothing. The activation of Ag nanoparticles and Zn nanowires in electrospun poly(methyl methacrylate) (PMMA) nanofibres has provided conventional garments with antibacterial, antiviral, and self-cleaning properties [25].

To date, most studies related to nanofibres have mainly focused on their functional properties and technical applications, leaving the aesthetic aspect of these nano-sized structures largely unexplored. While significant progress has been made in understanding and exploiting the unique properties of nanofibres and the other technical characteristics of various materials, it is surprising to note that the visual and aesthetic impact of these fibres has often taken a back seat. As nanofibres are increasingly integrated into various products and materials, it is essential to consider how they will visually affect the result. It is essential to recognise that aesthetics is not merely a superficial aspect but can influence the overall perception of a product's quality and usefulness.

The study of the aesthetic aspect of nanofibres involves understanding how they interact with light, how they influence colour perception, and how they can be modified to achieve desirable visual effects [26]. Regarding the interplay between light and materials, a single nanofibre is not visible to the naked eye or through an optical microscope because the obtained nanometer diameters (50–1000 nm) [27] are generally in the same wavelength magnitude range as visible light (400–700 nm) [28]. However, a single nanofibre is only visible with an electron microscope. While it is feasible to visually detect the existence of nanofibres through the naked eye or optical microscopy under circumstances where these nanofibres create a dense layer enveloping a substrate, the reality is that a substantial accumulation of nanofibres, when illuminated by white light, presents itself as a white, impenetrable, and uniformly even surface. This white colour occurs when a thick layer of nanofibres is produced due to the physical phenomenon of light scattering [26].

Recent advancements in nanofibre dyeing have enabled the enhancement of their aesthetic properties. Researchers have studied two methods of dyeing nanofibres. The first method involves the incorporation of the dye compound directly into the polymer solution before the electrospinning process [26,29,30]. As the fibres are generated from the solution and deposited on the collector structure, an increase in the deposition of nanofibres leads to an increase in the perceived colour intensity. This approach can achieve more intense and uniform colours from the outset due to the direct incorporation of dyes into the fibre structure itself. In contrast, the other method of dyeing nanofibres takes place after the electrospinning process, i.e., the fibres are immersed in a dye solution [31–33]. This allows more precise control over the intensity and uniformity of the colour, as the fibres can be immersed in the solution for different periods to achieve the desired shade.

In this work, we focused on composites made of a textile and coated with a nanofibre layer. The nanofibres were colour-treated by encapsulating dyes inside them using two different techniques: polymer colouration and liquid solution encapsulation. Two different polymer solutions were prepared for electrospinning, thus comprising two different types of dyeing. To obtain one solution, the dye was dispersed directly on an aqueous polymeric solution of polyvinyl alcohol (PVA). To prepare the other solution, the dye was solubilised in an oily solution of sage essential oil, which was then emulsified in a polymeric solution of PVA with a surfactant. PVA was selected as the polymer because it is a water-soluble polymer and is compatible with electrospinning. The two dyes were used, one water-soluble and the other oil-soluble, and were black in colour. This colour was selected because of its high contrast with the surface where the nanofibres were collected (which was white), as, in this way, the colour can be seen with the naked eye and with short electrospinning process times. The dyes were selected based on their molecular weights since their very similar values rule out their influence on the dyeing power of the nanofibres or the obtained colour. Sage essential oil was used because it has shown good levels of encapsulation in the nanofibres in previous studies as well as because it is an almost colourless oil, which facilitates the appreciation of the dyeing.

2. Materials and Methods

2.1. Materials

Polyvinyl alcohol (PVA) with a molecular weight (Mw) of 61,000 g/mol was obtained from Sigma-Aldrich (Akralab, Alicante, Spain) and used to create the solutions. Distilled water was utilised to prepare both solutions. As colouring compounds, the water-soluble dye Acid Black 194 (AB194) supplied by BASF (Tarragona, Spain) and the oil-soluble dye Solvent Black 3 (SB3) supplied by ANALEMA were used, its chemical structure is shown in Figure 1. The essential oil used in the experiment was sage, which was purchased from Lozano Essences (Esencias Lozano, S.L., Murcia, Spain). For surfactant purposes, Tween 80 from Panreac (Akralab, Alicante, Spain) was employed.

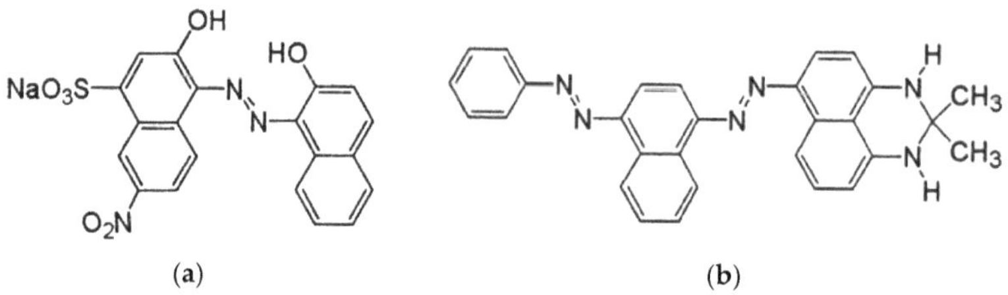

Figure 1. Chemical structures of the dyes. (**a**) Acid Black 194; (**b**) Solvent Black 3.

The nanofibres were deposited on a plain white weave fabric comprising 100% polyester (PES), 200 g/m^2.

2.2. Methods

2.2.1. Preparation of the Solutions

Both solutions were obtained by preparing a 9% (*w*/*v*) PVA solution. This involved heating the water to 80 °C via magnetic stirring until the PVA was completely dissolved. To prepare the solution with the Acid Black 194 dye, it was added to the PVA solution at a concentration of 5 g/L and left to stir for 2 h at room temperature in a propeller homogeniser. To prepare the emulsion, the essential oil was added to the PVA at room temperature. Sage oil was added at 3% (*v*/*v*) and Tween 80 surfactant at 1% (*v*/*v*) to the PVA solution. Initially, the PVA solution was placed in the homogeniser and when the desired speed was reached, the amount of Tween 80 was gradually added. Subsequently, a set amount of sage essential oil was gradually added. After the addition of oil was finished, the revolutions were sustained for 5 min. The homogeniser and its respective conditions were as follows: an Ultraturrax homogeniser with a toothed accessory at 4000 rpm for 10 min for the surfactant and 1500 rpm for 5 min for the sage oil.

2.2.2. Characterisation of the Solutions

The obtained solutions were characterised using a variety of methods. The viscosity of the two solutions was measured with the Visco Elite R viscometer (Fungilab, Sant Feliu de Llobregat, Spain). The selection of the measuring device was based on the manufacturer's guidelines for the specific measuring range achieved. For each solution, the conductivity was measured using a Crison Conductimeter Basic 30 (Hach Lange Spain, S.L.U., L' Hospitalet de Llobregat, Spain). Additionally, the surface tension was measured using a Krüss tensiometer K9 (Krüss, Hamburg, Germany). The pH was measured with the GLP 22 CRISON pH metre (CRISON, Barcelona, Spain). Three samples were prepared, and every sample was tested for each parameter (viscosity, conductivity, surface tension, and pH) five times.

2.2.3. Electrospinning Process

The electrospinning process was conducted using a BIOINICIA electrospinning system (Bioinicia, Paterna, Spain). A 100% polyester (PES) bleached fabric with a taffeta weave was placed on the vertical stainless steel flat collector, where the nanofibres were later collected. A 22-gauge extruder capillary was used for all electrospinning processes. The parameters of the electrospinning process are listed in Table 1, which were optimised to avoid beds [34].

Table 1. The electrospinning parameters.

Reference	PVA	PVA + Acid Black 194	PVA + Sage oil + Solvent Black 3
Flow Rate (mL/h)	0.3	0.45	0.35
Voltage (kV)	10.5	10	11.5
Nozzle–collector distance (cm)	15	15	15
Process time (min)	90	90	90
Room temperature (°C)	24	22.5	26
Room humidity (%)	44	57	40

The parameters given in Table 1 were selected to facilitate a constant and stable electrospinning process. A balance between the selected voltage and the flow rate is imperative; if an excessively low flow rate is applied relative to the applied voltage, electrospinning will be interrupted, whereas if a greater flow rate than the voltage can withstand is applied, an irregular Taylor cone will form and solution droplets will be deposited on the collector, not just the nanofibres. The nozzle–collector distance was selected so that the solvent used in the solutions evaporates in that space–time. All parameters have proven effective in previous research.

2.2.4. Characterisation of Nanofibres

The first characterisation of the nanofibres was performed to determine if there was visible colouring to the naked eye due to the encapsulation of the dye in each of them. Three different electrospun samples, PVA nanofibres, PVA + Acid Black 194 nanofibres, and PVA + sage oil + Solvent Black 3 nanofibres, were presented to five individuals who were asked to determine the appearance of colour in each sample referenced with letters of the alphabet, avoiding the power of suggestion on the volunteers' opinions.

Subsequently, in parallel with the opinions of the five individuals, reflection spectroscopy was carried out to determine the chromatic coordinates of the produced nanofibres. For this purpose, measurements were carried out with the Datacolor Spectro 700 reflection spectrophotometer; the measurement parameters were illuminant D65 and observer at 10°, including the specular component, and an aperture size of 3 cm. Analysis of the samples' colours was conducted via the CIEL*a*b* colour space, as it is the most comprehensive colour space. The three coordinates of CIELAB space represent the lightness of the colour ($L^* = 0$ indicates black and $L^* = 100$ indicates white), its position between red and green ($a^* < 0$ indicates green colour while $a^* > 0$ indicates red), and its position between yellow and blue ($b < 0$ indicates blue while $b < 0$ indicates yellow colour). The three veils were measured, and each one was measured in five different zones.

For the morphological characterisation, scanning electron microscopy (SEM) was employed via a FESEM ULTRA 55 (Carl Zeiss, Jena, Germany) with an accelerating voltage of 2 kV. The surfaces of each sample were analysed at appropriate magnifications for the study. Before observation, the samples were coated with gold/platinum to ensure the necessary conductivity.

Fourier transform infrared spectroscopy (FTIR) was conducted to characterise the starting materials (PVA, dyes, and sage oil) and the obtained nanofibres. A JASCO FT/IR-4700 type A spectrophotometer (Jasco Spain, Madrid, Spain) with an ATR accessory (Jasco Spain, Madrid, Spain) was used. Each sample underwent sixteen spectra measurements with a resolution of 4 cm^{-1}.

To ensure uniformity in size distribution measurements, we utilised the image analysis software Image J 1.52p (Wayne Rasband, MD, USA). Each image of the samples under analysis was accurately calibrated to obtain measurements in the correct units. The obtained measurements were then transferred to Excel, and corresponding representations were generated. Three different nanofibre veils were obtained, and 20 measures were made from every image.

Finally, to evaluate the staining capacity of the nanofibres after subjection to a 20-cycle rubbing process, the Crockmeter model CM-1 from ATLAS ELECTRIC DEVICES Co. (Chicago, IL, USA) was used. The control fabric used during the test was the same fabric on which the nanofibres were deposited, i.e., a 100% polyester fabric.

3. Results

3.1. Solution Characterisation

The characteristics of all the solutions and the components used are given in Table 2. All measurements were carried out at room temperature so as not to influence the parameters.

Table 2. Solution parameters.

Reference	PVA 9%	PVA + Acid Black 194	PVA + Sage Oil + Solvent Black 3
Viscosity (cP) (SD 3.47)	143.88	177.36	181.66
Conductivity (μS) (SD 6.78)	294.2	2620	248.8
Surface tension (mN/m) (SD 1.76)	59.58	54.24	33.86
pH (SD 0.03)	5.62	6.32	5.09

We correctly characterised the solutions by measuring the viscosity, surface tension, conductivity, and pH of each one, as well as the starting solution of 9% (*w*/*v*) PVA. We also measured the solution parameters of Solvent Black 3 dye and sage oil to determine their influence on the emulsion. Table 2 shows how the viscosity values increase with the addition of the colouring substances in the PVA solution. However, the opposite was true for surface tension values, which decreased. The conductivity parameter increased considerably when AB194 dye was added; however, it decreased slightly when SB3 dye and sage oil were added. The pH values remained partially stable.

Figure 2 shows the images of the solutions under an optical microscope. As expected, no compounds or aggregates were found in the PVA solution (Figure 2a), which means that the polymer was correctly dissolved. The same happened with the dissolution of the AB194 dye in the PVA solution (Figure 2b); a completely homogeneous solution was observed due to the great solubility of the dye in the aqueous solution. However, at first glance, it can be observed that the solution had lost its initial transparency attributed to the PVA and now had a high black colouring. Finally, due to the hydrophilic and hydrophobic nature of the components, an emulsion was obtained between PVA (aqueous) and sage essential oil dyed with Solvent Black 3 (oil). Figure 2c shows the microdroplets generated by the emulsion, which were characterised by a homogeneous distribution of the encapsulations' radial sizes.

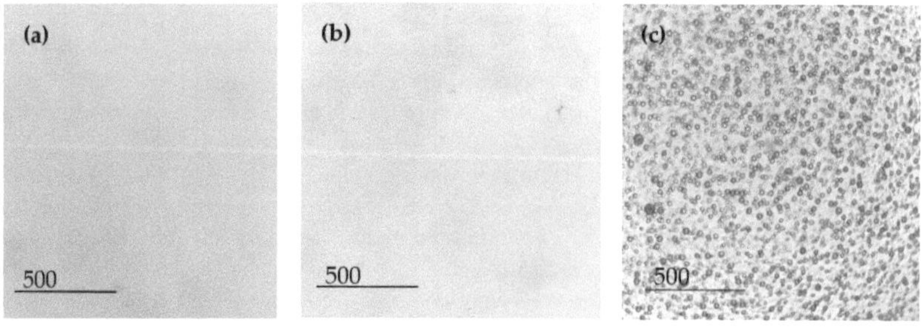

Figure 2. Different solutions under an optical microscope. (**a**) PVA solution; (**b**) PVA + Acid Black 194 solution; (**c**) PVA + sage oil + Solvent Black 3 emulsion.

Maintaining the stability of an emulsion over time is complicated, and this becomes even more difficult when voltage is applied to the solution. The electrospinning technique requires the creation of an electrostatic field between two electrodes by applying an electric current to the extruder component. Figure 3 shows the effect of voltage application on the stability of the emulsion.

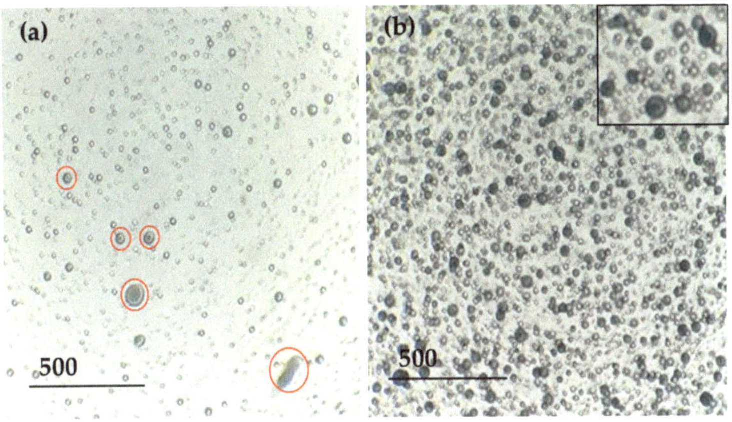

Figure 3. PVA + sage oil + Solvent Black 3 emulsion under an optical microscope. (**a**) Solution extracted from the extruder capillary where the electric current is applied, red circles show different size drops; (**b**) solution extracted from the syringe where the emulsion is held.

The application of electric current to an emulsion can separate its component phases; in this case, PVA is the aqueous phase and sage oil is the oily phase. For years, many researchers have shown interest in electrostatic phase separation for the separation of oil from water. When an electric current is applied to an emulsion, electrocoalescence can occur. This means that when an electric field is applied, the dispersed emulsion droplets start to move and experience electrostatic forces that can lead to coalescence until they merge. In addition, if the outer membrane of the droplet becomes too narrow, it can break and separate the dispersed and dispersing phases [35,36]. Figure 3a shows a big difference from the initial emulsion image; the encapsulated droplets were greatly reduced, and a different size distribution was generated. Figure 3a shows a big difference concerning the image of the initial emulsion. After applying electric current in the electrospinning process, the encapsulated droplets were greatly reduced, and a different size distribution was generated. In addition, it can be seen (Figure 3a, marked in red circles) that the encapsulation of the oil in the droplets has apparently decreased. However, as shown in Figure 3b, the emulsion that was retained in the syringe of the equipment was stable, just like the initial emulsion, and the encapsulated black-dyed sage oil could be seen (see Figure 4).

3.2. *Nanofibre Characterisation*

3.2.1. Visual Colour Test

The visual staining test was carried out with five volunteers who were shown three samples and asked to answer which one had black staining. Figure 5 shows the electrospun dye nanofibres.

As can be seen from the survey results (Table 3), all respondents were able to identify the colour black in only one sample; the other two apparently showed a white colour.

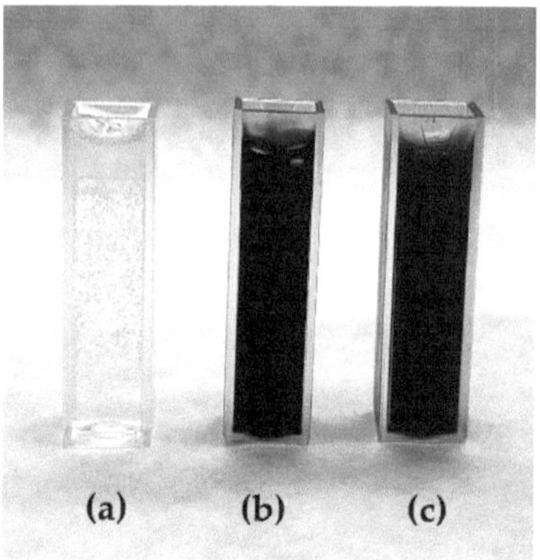

Figure 4. Solutions used in the experiment. (**a**) PVA 9%; (**b**) PVA + Acid Black 194; (**c**) PVA + sage oil + Solvent Black 3.

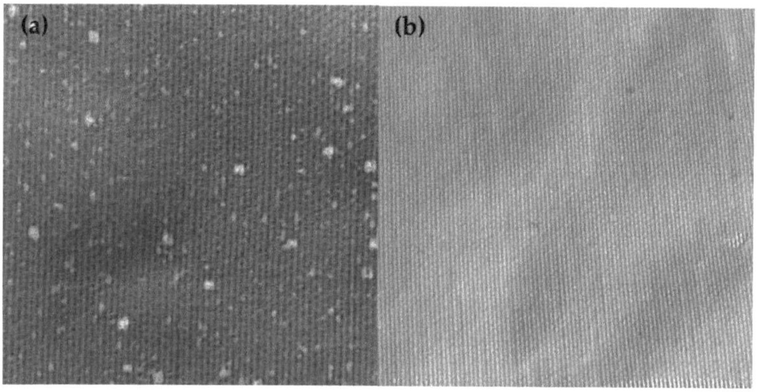

Figure 5. Nanofibres on polyester fabric. (**a**) PVA + Acid Black 194 nanofibres; (**b**) PVA + sage oil + Solvent Black 3 nanofibres.

Table 3. Visual colour test results.

Reference	PVA 9%	PVA + Acid Black 194	PVA + Sage Oil + Solvent Black 3
Colour Identification (five volunteers)	White (5)	Black (5)	White (5)

3.2.2. Diffuse Reflectance Spectroscopy

Diffuse reflectance spectroscopy was used to perform a detailed and quantitative analysis of the obtained colour data. The chromatic coordinates of the samples were determined by averaging at least four measurements of the samples.

Table 4 shows the CIE Lab* colourimetry values obtained for each sample. The 100% PES textile was used as the standard sample because the nanofibres were electrospun on it.

Figure 6 shows the location of the values obtained in CIE Lab* space. The values obtained for the PVA nanofibres (yellow point) showed minimal colour variation with respect to the white polyester fabric. This could be because, although the PVA solution is transparent in its liquid state, when it solidifies, it turns white or slightly yellow. The brightness of the sample was very similar to the standard sample.

Table 4. CIE Lab* data of the samples.

Reference	L* (SD 0.81)	a* (SD 0.19)	b* (SD 0.21)	dL*	da*	db*	dE*ab	Difference
PES (Standard)	95.29	−0.89	1.91	-	-	-	-	-
PVA 9%	94.83	−0.87	2.12	−0.46	0.02	0.21	0.50	No
PVA + Acid Black 194	76.39	−0.75	−1.73	−18.89	0.14	−3.64	19.24	Yes
PVA + sage oil + Solvent Black 3	94.99	−0.88	2.07	−0.3	0.01	0.16	0.34	No

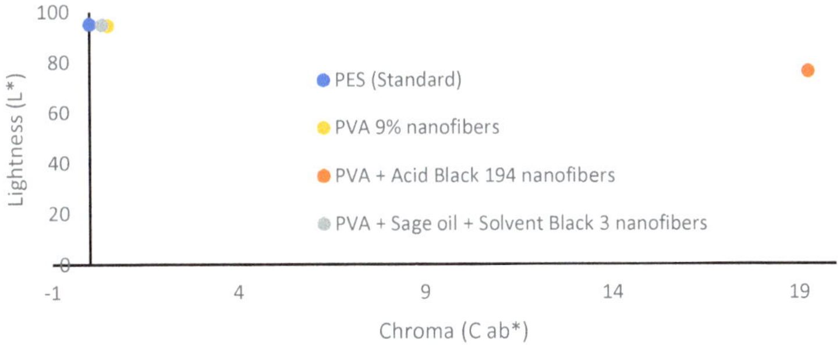

(**a**)

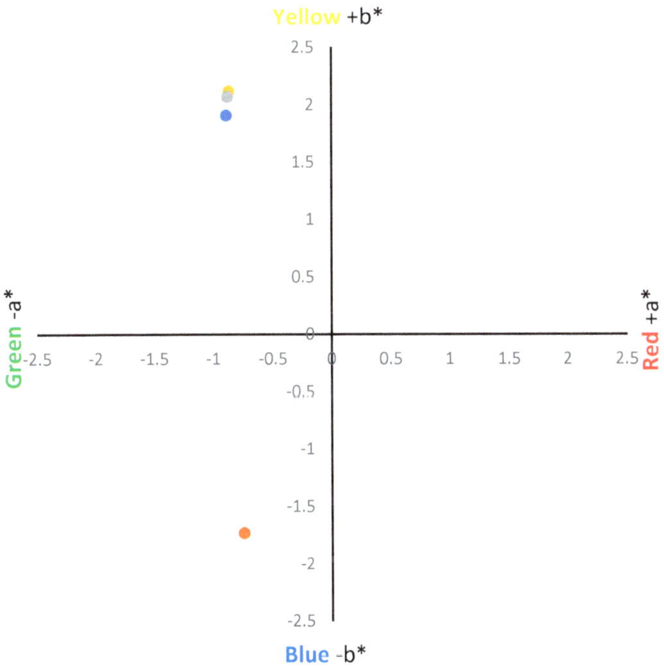

(**b**)

Figure 6. Results of the colourimetry of the samples. (**a**) Sample brightness results. (**b**) Chromatic coordinates: blue point: PES fabric (standard); yellow point: PVA 9% nanofibres; orange point: PVA + Acid Black 194 nanofibres; grey point: PVA + sage oil + Solvent Black 3 nanofibres.

On the contrary, the results showed that the addition of Acid Black 194 dye to the PVA solution generated bluish-coloured nanofibres, as indicated by its value (−1.73 b*), which can be seen in the position of the orange point in Figure 6b. This result is compatible with what the five volunteers answered in the visual test. It also decreased the brightness value (76.39 L*) of the sample compared to the standard (95.29 L*) due to dyeing.

However, although the emulsion of PVA + sage oil + black solvent 3 had a greyish colour in the liquid state, this colour did not appear on the nanofibres when they were electrospun. The CIE Lab* measurements showed a value of −0.88 for parameter a* and a value of 2.07 for b* for the nanofibres generated from the PVA + sage oil + Solvent Black 3 emulsion. These values are very close to those obtained for the standard sample (−0.89 a* and 1.91 b*), which indicate that the nanofibres generated from the emulsion did not show a black colouring on their surface but showed a slightly yellowish–whitish shade, similar to the PVA nanofibres.

3.2.3. Rubbing Test

We sought to determine what happens to both types of nanofibres when subjected to a rubbing process. After the test, the colours of the rubbed sample and the control used for the rubbing were evaluated. Table 5 shows the colourimetry results obtained after rubbing.

Table 5. CIE Lab* data of the samples after the rubbing test.

Reference	L* (SD 0.76)	a* (SD 0.17)	b* (SD 0.18)	DL*	Da*	Db*	DE*ab	Difference
PES (Standard)	95.29	−0.89	1.91	-	-	-	-	-
PVA + Acid Black 194_Rubbed	91.45	−0.94	0.78	−3.84	−0.05	−1.14	4.01	Yes
PVA + Acid Black 194_Control	76.91	−0.64	−2.7	−18.38	0.25	−4.62	18.95	Yes
PVA + sage oil + Solvent Black 3_Rubbed	94.24	−0.78	2.11	−1.05	0.11	0.19	1.07	Yes
PVA + sage oil + Solvent Black 3_Control	93.41	−0.82	2.06	−1.88	0.07	0.14	1.89	Yes

Figure 7 shows the brightness results obtained after the rubbing tests. Concerning the nanofibres containing the AB194 dye (PVA + Acid Black 194), it can be seen that after performing the test on the electrospun sample (PVA + Acid Black 194_Rubbed), the brightness value of the latter increased to 91.45 L*, a value that almost reached that of the standard PES sample, 95.29 L*. In the same way, it can be seen that the brightness value of the control fabric (PVA + Acid Black 194_Control), which would initially be 95.29 L* as it is the same standard fabric, decreased to 76.91 L*, a value very similar to that of the PVA + Acid Black 194 sample, 76.39 L*. These results can be explained by the fact that the rubbing process caused the nanofibres to detach from the PES fabric, where they were deposited and transferred to the used control fabric.

However, when analysing the brightness values of the rubbed sample with the encapsulated SB3 dye (PVA + sage oil + Solvent Black 3_Rubbed) and the control fabric (PVA + swage oil + Solvent Black 3_Control), these values are not very different from those obtained with the original sample.

Figure 8 shows the colour difference results obtained after the rubbing tests.

The colour difference data obtained corroborate the abovementioned; in the case of the PVA + Acid Black 194 sample, the dyed nanofibres or dye was directly transferred to the control fabric (PVA + Acid Black 194_Control) used. For this reason, the control, which was initially white after the test, had a DE*ab value of 18.95. As the value increased in the control sample, it decreased in the rubbed sample, as it obtained a value of 4.01 DE*ab.

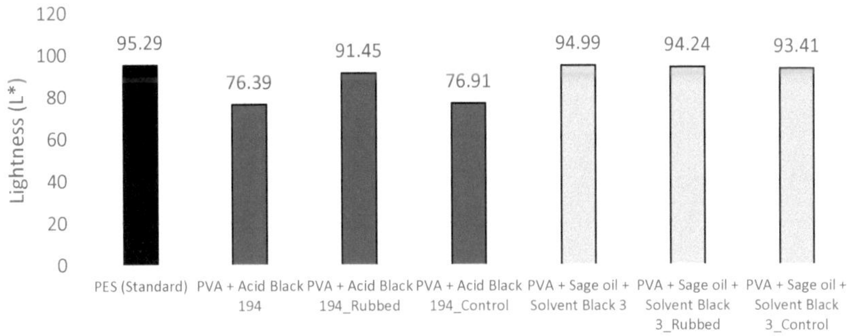

Figure 7. Luminescence (L*) results of the samples after the rubbing test.

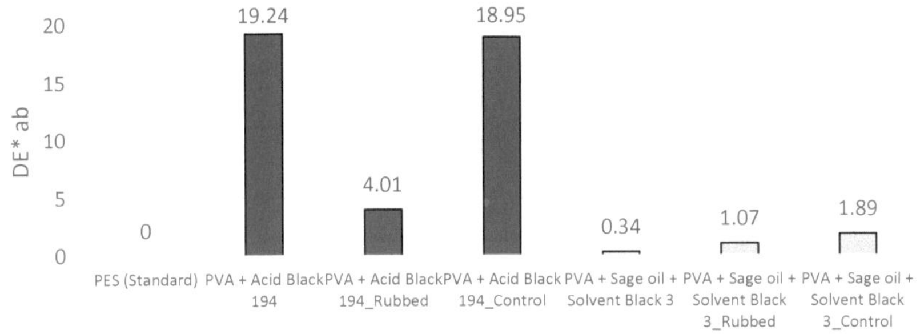

Figure 8. Results of colour difference (DE*ab) of the samples after the rubbing test.

After the rubbing of the PVA + sage oil + Solvent Black 3 sample, the DE*ab value increased slightly (1.07 DE*ab) compared to the original sample (0.34 DE*ab). However, a higher increase in the control sample was observed, which reached a value of 1.89 DE*ab.

Despite the rubbing tests, a greater colour difference was still observed in the samples containing the Acid Black 194 dye compared with the nanofibres produced from the emulsion with sage oil and Solvent Black 3 dye.

3.2.4. FTIR

Fourier transform infrared spectroscopy (FTIR) is highly valuable in discerning functional groups exhibiting distinct vibrations within specific spectral ranges, typically between 4000 and 400 cm^{-1}. Nonetheless, achieving quantitative determinations demands meticulous calibrations and involves increased complexity due to the overlapping molecular vibrations in certain regions of the spectrum [37]. These overlaps can lead to fluctuations in the centre of the vibrational bands, making accurate measurements more challenging. In this case, we intended to use FTIR to identify the presence of the different dyes and the essential oil of sage in the nanofibres so as to identify the functional groups characteristic of the type of dye and the essential oil used and determine the evolution of the PVA curve when the additive compounds are included.

When analysing the spectrum of the nanofibres obtained from the PVA + Acid Black 194 solution and comparing it with the spectra of the PVA and the dye, the presence of peaks characteristic of both can be seen in Figure 9. This graphic shows the polyvinyl alcohol nanofibre in blue, the spectrum of the AB194 dye in orange, and the spectrum of the PVA + Acid Black 194 in grey.

The characteristic peaks, named A, B, and C, have already been cited by several authors who have analysed the same or similar compounds. Point A marks a characteristic peak of PVA at wavelength 3300 cm^{-1}; this is due to its -OH groups [38]. The characteristic peak of the dye called B at 1580 cm^{-1} is due to the presence of the azo bond (-N=N-), characteristic of azo dyes [39,40]. Finally, the C peak at wavelength 1100 cm^{-1} shows the aromatic nature of the dye, i.e., it reflects the presence of carbon–carbon bonds (C=C) [40,41].

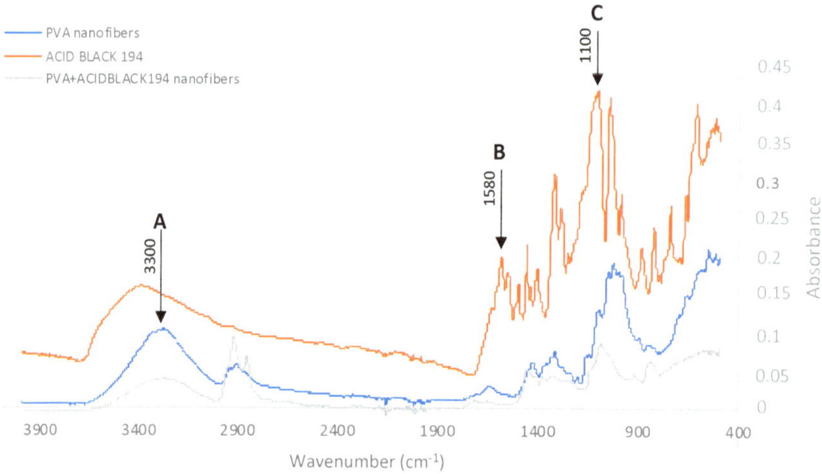

(**a**)

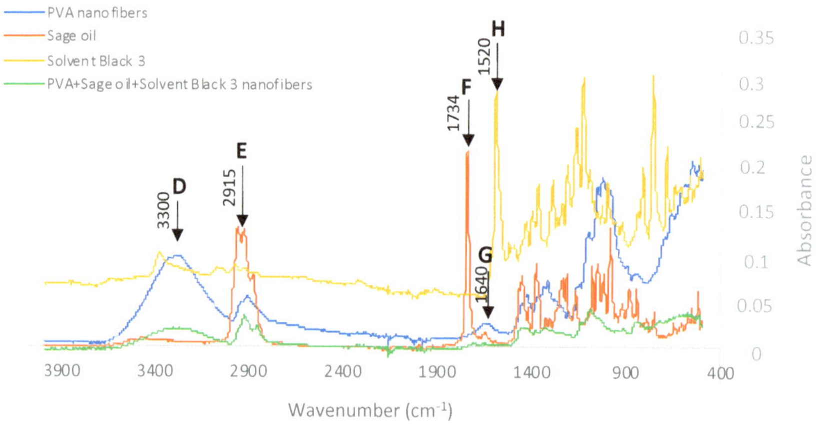

(**b**)

Figure 9. Infrared spectra of the nanofibres. (**a**) Acid Black 194; blue line: PVA nanofibres; orange line: Acid Black 194 dye; grey line: PVA+Acid Black 194 dye nanofibres. (**b**) Solvent Black 3; blue line: PVA nanofibres; orange line: sage oil; yellow line: Solvent Black 3 dye; green line: PVA + sage oil + Solvent Black 3 dye nanofibres.

The presence of the AB194 dye should provide an increase in the -N=N- band with respect to the OH band. This is reflected in Table 6, where the spectrum of PVA nanofibres, the spectrum of Acid Black 194, and the spectrum of the PVA+ Acid Black 194 nanofibre veil are analysed.

The intensity analysis showed that the presence of AB194 in the PVA nanofibres increased the ratio (I_{1580}/I_{3300} = 1.3516) with respect to that of the nanofibres without dye (I_{1580}/I_{3300} = 0.1650). The opposite occurred when the C=C band was studied with respect to the OH band. In this case, the presence of the dye could be seen when the absorbance of the coloured nanofibres (I_{3300}/I_{3100} = 0.4684) decreased with respect to the PVA nanofibres (I_{3300}/I_{3100} = 0.7946). In this way, it is possible to demonstrate the presence of Acid Black 194 in the PVA nanofibres, even if it is already visually observed.

Table 6. Band intensity of FTIR spectra of electrospun PVA nanofibres, Acid Black 194 dye, and electrospun PVA+ Acid Black 194 nanofibres.

Reference	A I_{3300}	B I_{1580}	B/A I_{1580}/I_{3300}	C I_{1100}	A/C I_{3300}/I_{1100}
PVA nanofibres	0.3435	0.0772	0.1650	0.0618	0.7946
ACID BLACK 194	0.1591	0.1922	1.2080	0.4142	0.3841
PVA + Acid Black 194 nanofibres	0.0364	0.0492	1.3516	0.0777	0.4684

Similarly, to determine the presence of sage oil and SB3 dye in the electrospun PVA nanofibres, their FTIR spectra were analysed and are shown in Figure 9b. In this case, the remarkable peaks were named D, E, F, G, and H.

The peak centred around point D aligned with the peaks at 3300 cm^{-1}, primarily indicating OH stretching. This peak was notably absent in sage oil. Adjacent to this feature, we observed another peak labelled E, aligning with the range of 2956–2849 cm^{-1}. This range is attributed to both symmetrical and asymmetrical stretching of CH bonds (in CH3 and CH2 groups) [42]. Notably, this band exhibited significant intensity even when tested with sage essential oil. Consequently, the presence of sage can be recognised by an elevated CH band in comparison to the OH band. This observation is evident in Table 7, which displays the spectra of three samples: electrospun PVA nanofibres, the sage essential oil spectrum provided by the supplier, the spectrum of the Solvent Black 3 dye, and the spectrum of the PVA + sage oil + Solvent Black 3 nanofibres. The intensity assessment revealed that the presence of sage led to an augmentation in the intensity of the CH stretching band (I_{2915}) in relation to the OH stretching band (I_{3300}). As a result, the calculation of the ratio between these two bands (I_{3300}/I_{2915}) exhibited a substantial contrast between PVA (1.9902) and sage (0.0450). This discrepancy signifies that the presence of sage in sample PVA + sage oil + Solvent Black 3 nanofibres (0.6412) was confirmed through a reduction in this ratio compared to PVA devoid of sage.

Table 7. Band intensity of the FTIR spectra of electrospun PVA nanofibres, sage oil, and PVA+ sage oil + Solvent Black 3 nanofibres.

Reference	D I_{3300}	E I_{2915}	D/E I_{3300}/I_{2915}	F I_{1734}	G I_{1640}	G/F I_{1640}/I_{1734}	H I_{1520}	D/H I_{3300}/I_{1520}
PVA nanofibres	0.1017	0.0511	1.9902	0.0139	0.0259	1.8633	0.0212	4.7972
Sage oil	0.0058	0.1288	0.0450	0.2193	0.0143	0.0652	0.0040	1.4500
Solvent Black 3	0.0871	0.0849	1.0259	0.0597	0.0570	0.9547	0.2846	0.3092
PVA + sage oil + Solvent Black 3 nanofibres	0.0202	0.0315	0.6412	0.0027	0.0037	1.3703	0.0022	9.1818

Likewise, the distinct characteristics of salvia, namely camphor and thujone, displayed C=O stretching vibrations with a central frequency of 1730 cm^{-1}. Additionally, the presence of pinene was identifiable by the -C=C- alkene bond, as evidenced by a peak centred at 1640 cm^{-1} [43,44], which corresponds to typical features of terpenes found in essential oils [45]. In the spectra of the nanofibres (Figure 9b), these specific peaks aligned with the designated F and G peaks of the sage oil spectrum. Upon scrutinising the behaviour of these bands within the sage sample utilised in the study, it was evident that the presence of pinene was notably less prevalent in comparison to camphor and thujone. This was reflected by a distinctive peak at 1730 cm^{-1}. The data presented in Table 7 demonstrate that PVA nanofibres with sage led to a reduction in the ratio (I_{1640}/I_{1734} = 1.3703) when contrasted with PVA without sage (I_{1640}/I_{1734} = 1.8633). Consequently, the intensities of these bands provided further substantiation for the existence of sage within the nanofibres.

To identify the Solvent Black 3 dye inside the nanofibres, the so-called H peak (Figure 9b) was observed at wavelength 1520 cm-1 due to the stretching of the azo groups (N=N) [46]. By studying the ratios in Table 7, it can be seen that with the addition of the SB3 dye, the ratio (I_{3300}/I_{1520} = 9.1818) increased with respect to that observed when the nanofibres were produced without dye (I_{3300}/I_{1520} = 4.7972), which may be evidence of the presence of dye inside the PVA and sage oil nanofibres.

3.2.5. SEM

SEM characterisation was performed to assess whether there was any difference at the nanoscale according to the electrospinning solution. Figure 10a shows the nanofibres generated from the dissolution of AB194 dye in the PVA solution. It can be seen that the aqueous solvent evaporated correctly during the electrospinning process, as the nanofibres did not have pores on their surfaces. The nanofibres were mostly tubular in cross-section, but there were many elongated beads of similar size along them. Due to the long process time of 90 min, a high deposition of the nanofibres and the creation of different nanofibrous layers on the PES collector substrate were observed.

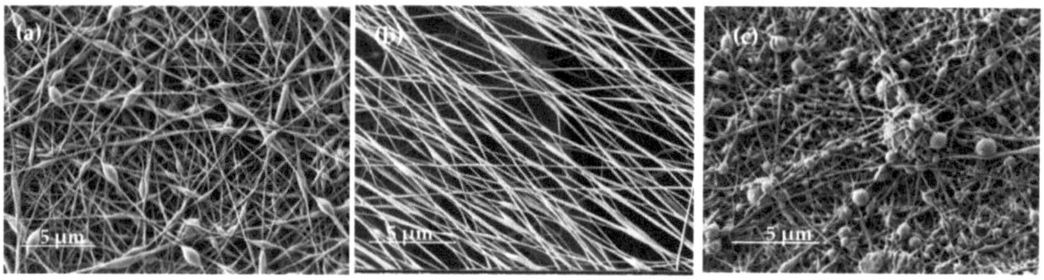

Figure 10. SEM images of the nanofibres. (**a**) PVA + Acid Black 194 nanofibres (5.00 kx) in bad condition (with beds); (**b**) PVA + Acid Black 194 nanofibres (5.00 kx); (**c**) PVA + sage oil + Solvent Black 3 nanofibres (5.00 kx).

On the other hand, when the solution prepared from the emulsion of sage oil dyed with SB3 dye in an aqueous solution of PVA was electrospun, a different result was obtained. Figure 10b shows the produced nanofibres, which had more beads in their cross-section; in this case, the beads were also more spherical in shape. Previous studies have shown that the spherical beads contain the sage oil used in the emulsion [34]. In both cases, it is observed that the deposition of the nanofibres was completely random due to the use of a static collector. Table 8 shows the average diameter of the nanofibres and beads.

Table 8. Average diameter of the nanofibres and beads.

Reference	PVA + Acid Black 194	PVA + Sage Oil + Solvent Black 3
Ø Nanofibres (nm) (SD 2.81)	87.83	68.94
Ø Beads (nm) (SD 1.99)	271.04	430.20

Bead formation in nanofibre veils can be due to multiple factors. However, it is generally attributed to an unstable Taylor cone due to an inexact balance between the feed flow rate and applied voltage. A lower viscosity favours the appearance of larger-diameter beads. When Acid Black was added to 9% PVA, as shown in Table 2, the viscosity slightly increased from 143.88 cP to 177.36; the flow rate and voltage need to be adjusted as described in Table 1; otherwise, beads appear (Figure 10a). Once the parameters were adapted to the solution's viscosity, no beads were present in the PVA + Acid Black 194 nanofibres, as can be seen in Figure 10b. The nanofibres showed a diameter of around 87.3 nm, and slight widening was observed, which is attributed to the dyestuff presence in the polymer solution, whose size is 271.04 nm. In contrast, the PVA + sage oil + Solvent Black 3 solution had a higher viscosity, 181.66 cP, so the beads formed in the nanofibres were of a higher diameter, 430.20 cP. It can be seen that the beads showed a spherical shape, which is likely due to the round shapes that keep the oil inside [34]. When the nanofibres were measured, it can be easily appreciated (Table 8) that the fineness decreased considerably when the 9% PVA was electrospun from the oil emulsion (68.94 nm) in comparison to the 9% PVA with dyestuff (87.83). This can be due to the stretching generated by the oil sphere in the polymer in order to be fully coated with PVA.

SEM images of the nanofibrous samples that were subjected to a rubbing test demonstrate the non-existence of nanofibres on the polyester fabric after the test (Figure 11a,c). This also demonstrates the low adhesion of the dyed PVA nanofibres to the collector substrate.

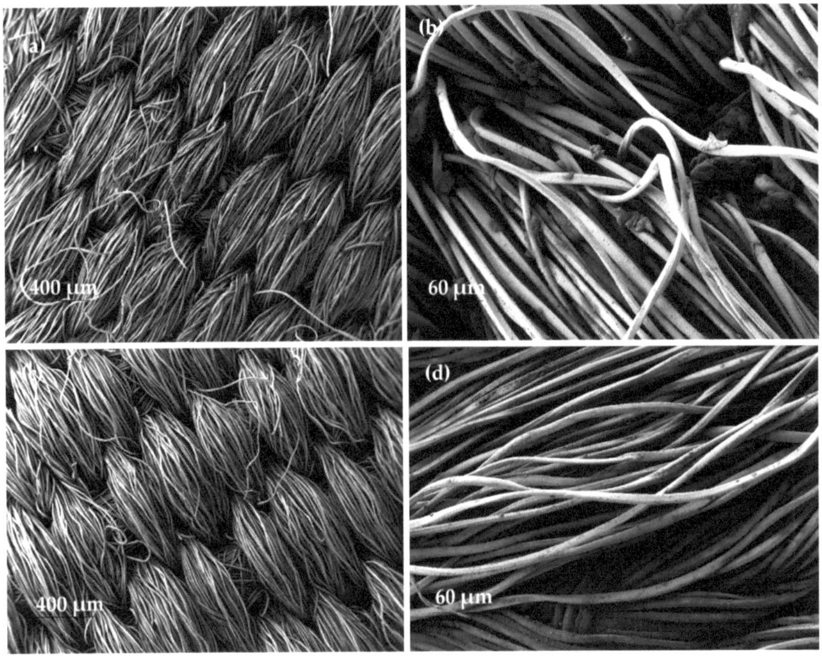

Figure 11. SEM images of the nanofibres after the rubbing test. (**a**) PVA + Acid Black 194 nanofibres (24 ×); (**b**) control fabric for the rubbing test on PVA + sage oil + Solvent Black 3 nanofibre sample (200 ×); (**c**) PVA + sage oil + Solvent Black 3 nanofibre (24 ×); (**d**) control fabric for the rubbing test on PVA + sage oil + Solvent Black 3 nanofibre sample (200 ×).

In addition, on the control fabric used to rub the nanofibres, a multitude of small bulges appeared along the polyester fibres and between them (Figure 11b,d). These bulges were not present on the original fabric before the test was carried out.

4. Discussion

In the electrospinning technique, the use of different compounds can significantly alter the morphology of the nanofibres. The initial characterisation of the two studied solutions shows how the initial values of the PVA solution can be modified by adding various compounds. Nevertheless, very similar electrospinning process values were established for both processes.

The optical microscopy performed on the polymeric solutions demonstrated the difference between dissolving the AB194 dye in an aqueous solution of PVA, where the obtained image was very similar to that of the PVA solution compared with the image obtained from the emulsion of the RB3 dye dissolved in sage essential oil together with the

PVA solution. The image of the emulsion showed the microcapsules generated where the coloured oil was stored due to its darker appearance.

The 5 g/L concentration of dye in the PVA solution was shown to be able to colour the nanofibres and be visible to the naked eye. However, when this amount was added to the sage oil and the emulsion was prepared, the produced nanofibres showed no noticeable colouring.

A quantitative analysis of the colour of the nanofibres by means of diffuse reflectance spectroscopy enabled us to establish the exact colour differences between the samples. The PVA and PVA + sage oil + Solvent Black 3 nanofibres presented brightness values very close to the values obtained by the collector substrate where the nanofibres were deposited, i.e., 94.83 and 94.99 L*, respectively. However, the PVA + Acid Black 194 sample presented a lower brightness value than the rest, 76.39 L*, due to the dark colouring obtained. In addition, the latter sample shifted on the *X*-axis shown in Figure 6a, which is explained by the large colour difference in the sample compared with the standard polyester collector sample.

The nanofibre veils produced from the solution of AB194 in the PVA solution had a greyish colour, as can be seen in the lower-left grid of Figure 6b, closer to the blue colour due to its value of −1.73 b*. In the case of the nanofibrous veils produced from the emulsion of sage oil dyed with the SB3 dye in the PVA solution, the location of the point that represents it in the graph is very close to the representative of the PVA nanofibres and the standard polyester fabric, which reflects the non-existence of surface colour in the generated nanofibres.

Rubbing tests are useful to determine what happens to the dyed nanofibres when the fabric is rubbed. The results on the PVA + Acid Black 194 sample showed that when the fabric with nanofibres was subjected to a rubbing test, the brightness of the fabric increased (91.45 L*). This could be attributed to the removal of the nanofibres from the fabric, as this results in a value very similar to that obtained on the collector fabric (95.29 L*). When analysing the control fabric with which the test was carried out, which initially had a value of 95.29 L* as it was the same fabric, this value dropped to 76.91 L*. This decrease could mean that a large part of the nanofibres that were initially on the sample to be tested were transferred to this control fabric. On the PVA + sage oil + Solvent Black 3 sample, no significant changes were observed after the rubbing test.

The previously assumed transfer of the nanofibres to the control fabric was confirmed by the evaluation of the colour difference shown in Figure 8. The obtained values showed a large decrease in the value of the rubbed PVA + Acid Black 194 sample (4.01 DE*ab) compared with the initial value (19.24 DE*ab), which reflects the disappearance of the nanofibres from the surface. Similarly, the value of the control sample increased to 18.95 DE*ab, which means that it can be categorically stated that almost all the nanofibres were transferred to the fabric with which the rubbing test was carried out. In the case of the nanofibres produced from the emulsion, it is worth noting that the values increased with respect to their original values after the rubbing test. The rubbed fabric reached a value of 1.07 DE*ab, a higher value than the 0.34 DE*ab of the original sample. The value of the control sample was 1.89 DE*ab. Therefore, as with the other solution, greater differences in colour were obtained in the rubbed sample because the nanofibres remained adhered to it.

The colourimetry results after the rubbing test on the nanofibres produced from the PVA + sage oil + Solvent Black 3 emulsion confirm the presence of encapsulated dye inside the nanofibres, which comes out when the polymeric shell from the nanofibre is destroyed. For this reason, the nanofibre veil generated from the emulsion looks very similar to the one produced from the PVA solution, as the SB3 dye remained encapsulated inside the fibre and only PVA was present on the outside.

FTIR characterisation was useful in demonstrating the presence of the Acid Black 194 and Solvent Black 3 dyes, as well as a sage essential oil, in the electrospun nanofibres.

SEM microscopy was able to demonstrate that nanofibres can be generated from both solutions and also provided information regarding the morphology of the fibres, highlighting the appearance of elongated beads along the tubular section of the nanofibres electrospun from PVA and AB194. In this case, an average value of 87.83 nm was obtained for the tubular diameter of the nanofibres and an average diameter of 471.04 nm for the beads. SEM images showed the difference when electrospinning the SB3 sage oil emulsion with PVA. These showed a greater number of beads along the nanofibres with a much more spherical shape. The beads had an average diameter of 430.20 nm, while the nanofibres had an average diameter of 68.94 nm, both values being lower than those obtained from the PVA and AB194 solution.

In agreement with diffuse reflectance spectroscopy, SEM images also showed that the nanofibre veils disappeared in the scrubbed sample and were transferred as aggregates to the test fabric.

5. Conclusions

Nanofibres have been extensively studied in terms of their technical aspects; however, the aesthetics of nanofibres have hardly been of interest. In recent decades, nanofibres have become integrated into a multitude of sectors due to their specific characteristics. However, in some of them, such as fashion and biomedicine, the aesthetics of these nanometric structures are beginning to be important.

Conventional textile dyeing can be carried out in multiple ways; however, the dyeing of electrospun nanofibrous structures has not been extensively studied. In this study, the influence of producing electrospun nanofibres from two dyed solutions, one from a dye solution and the other from an emulsion, including oil-soluble dyestuffs, was studied. The results proved the ability to incorporate a dye compound by both processes and its correct subsequent electrospinning process with slight modifications in comparison to the polymer electrospun without dyestuff. Furthermore, although both solutions were initially quite similar in colour, a large difference in colour was evident when the nanofibres were obtained.

The results showed a higher colouring of the nanofibres produced from the PVA+ Acid Black 194 solution. Therefore, we conclude that if a product requires visibly coloured nanofibres, dissolving the dye directly in the PVA solution is the best solution.

On the other hand, the electrospinning of the Solvent Black 3 dye emulsion in the essential oil of sage and PVA showed a non-existent surface colouring due to the fact that the dye remained encapsulated inside the nanofibre, so a completely white surface was obtained on the outside. Subsequently, when this fibre was destroyed, the dye was visible on the surface. This result is ideal for those applications where an initial surface colouring modification is not required, and it is necessary to provide evidence when some stimulus comes into contact with the surface and consequently touches nanofibres and destroys them.

In addition, rubbing tests showed the disappearance of the colour when rubbing the collector substrate where the nanofibres were collected, which is evidence of the weak adhesion between the nanofibres and the polyester fabric. The low adhesion between the collector fabric and the electrospun nanofibres can open up different fields of application, but it can also prevent its application in other fields where a stronger bonding of the two structures is required.

This study aimed to open up new fields of application where aesthetics is important or a touch sensor is necessary, based on the two types of nanofibre colourings studied. These conclusions are limited to 9% PVA solutions and the dyestuff tested. Further studies with different polymers, dyestuffs, and surfaces should be conducted to generalise this behaviour.

Author Contributions: Conceptualisation, M.B.-A. and P.D.-G.; methodology, M.B.-A.; software, D.M.-G.; validation, J.G.-P. and P.D.-G.; formal analysis, D.M.-G. and M.B.-A.; investigation, D.M.-G.; resources, P.D.-G.; data curation, M.B.-A. and D.M.-G.; writing—original draft preparation, D.M.-G.; writing—review and editing, M.B.-A. and J.G.-P.; visualisation, D.M.-G.; supervision, M.B.-A.;

project administration, P.D.-G.; funding acquisition, P.D.-G. All authors have read and agreed to the published version of the manuscript.

Funding: This research was funded by the INSTITUTO VALENCIANO DE COMPETITIVIDAD EMPRESARIAL (IVACE) and the nominative line in favour of the Universitat Politècnica de València, approved by the Budget Law of the Generalitat Valenciana in the year 2023.

Data Availability Statement: The data presented in this study is available on request from the corresponding author.

Conflicts of Interest: The authors declare no conflict of interest.

References

1. Fadil, F.; Affandi ND, N.; Misnon, M.I.; Bonnia, N.N.; Harun, A.M.; Alam, M.K. Review on electrospun nanofiber-applied products. *Polymers* **2021**, *13*, 2087. [CrossRef] [PubMed]
2. Lyu, C.; Zhao, P.; Xie, J.; Dong, S.; Liu, J.; Rao, C.; Fu, J. Electrospinning of nanofibrous membrane and its applications in air filtration: A review. *Nanomaterials* **2021**, *11*, 1501. [CrossRef] [PubMed]
3. Lv, D.; Zhu, M.; Jiang, Z.; Jiang, S.; Zhang, Q.; Xiong, R.; Huang, C. Green electrospun nanofibers and their application in air filtration. *Macromol. Mater. Eng.* **2018**, *303*, 1800336. [CrossRef]
4. Zhu, M.; Han, J.; Wang, F.; Shao, W.; Xiong, R.; Zhang, Q.; Pan, H.; Yang, Y.; Samal, S.K.; Zhang, F.; et al. Electrospun nanofibers membranes for effective air filtration. *Macromol. Mater. Eng.* **2017**, *302*, 1600353. [CrossRef]
5. El-Aswar, E.I.; Ramadan, H.; Elkik, H.; Taha, A.G. A comprehensive review on preparation, functionalization and recent applications of nanofiber membranes in wastewater treatment. *J. Environ. Manag.* **2022**, *301*, 113908. [CrossRef]
6. Cui, J.; Li, F.; Wang, Y.; Zhang, Q.; Ma, W.; Huang, C. Electrospun nanofiber membranes for wastewater treatment applications. *Sep. Purif. Technol.* **2020**, *250*, 117116. [CrossRef]
7. Chen, H.; Huang, M.; Liu, Y.; Meng, L.; Ma, M. Functionalized electrospun nanofiber membranes for water treatment: A review. *Sci. Total Environ.* **2020**, *739*, 139944. [CrossRef]
8. Eatemadi, A.; Daraee, H.; Zarghami, N.; Melat Yar, H.; Akbarzadeh, A. Nanofiber: Synthesis and biomedical applications. *Artif. Cells Nanomed. Biotechnol.* **2016**, *44*, 111–121. [CrossRef]
9. Hu, X.; Liu, S.; Zhou, G.; Huang, Y.; Xie, Z.; Jing, X. Electrospinning of polymeric nanofibers for drug delivery applications. *J. Control. Release* **2014**, *185*, 12–21. [CrossRef]
10. Rasouli, R.; Barhoum, A.; Bechelany, M.; Dufresne, A. Nanofibers for biomedical and healthcare applications. *Macromol. Biosci.* **2019**, *19*, 1800256. [CrossRef]
11. Liu, Y.; Hao, M.; Chen, Z.; Liu, L.; Liu, Y.; Yang, W.; Ramakrishna, S. A review on recent advances in application of electrospun nanofiber materials as biosensors. *Curr. Opin. Biomed. Eng.* **2020**, *13*, 174–189. [CrossRef]
12. Moon, J.; Park, J.A.; Lee, S.J.; Zyung, T.; Kim, I.D. Pd-doped TiO_2 nanofiber networks for gas sensor applications. *Sens. Actuators B Chem.* **2010**, *149*, 301–305. [CrossRef]
13. Elkasaby, M.A.; Utkarsh, U.; Syed, N.A.; Rizvi, G.; Mohany, A.; Pop-Iliev, R. Evaluation of electro-spun polymeric nanofibers for sound absorption applications. *AIP Conf. Proc.* **2020**, *2205*, 020042. [CrossRef]
14. Rabbi, A.; Bahrambeygi, H.; Shoushtari, A.M.; Nasouri, K. Incorporation of nanofiber layers in nonwoven materials for improving their acoustic properties. *J. Eng. Fibers Fabr.* **2013**, *8*, 155892501300800412. [CrossRef]
15. Sinha, M.K.; Das, B.R.; Prasad, N.; Kishore, B.; Kumar, K. Exploration of nanofibrous coated webs for chemical and biological protection. *Zaštita Mater.* **2018**, *59*, 189–198. [CrossRef]
16. Dhineshbabu, N.R.; Karunakaran, G.; Suriyaprabha, R.; Manivasakan, P.; Rajendran, V. Electrospun MgO/Nylon 6 hybrid nanofibers for protective clothing. *Nano-Micro Lett.* **2014**, *6*, 46–54. [CrossRef]
17. Moreira, J.B.; de Morais, M.G.; de Morais, E.G.; da Silva Vaz, B.; Costa, J.A.V. Electrospun polymeric nanofibers in food packaging. In *Impact of Nanoscience in the Food Industry*; Academic Press: Cambridge, MA, USA, 2018; pp. 387–417. [CrossRef]
18. Amna, T.; Yang, J.; Ryu, K.S.; Hwang, I.H. Electrospun antimicrobial hybrid mats: Innovative packaging material for meat and meat-products. *J. Food Sci. Technol.* **2015**, *52*, 4600–4606. [CrossRef]
19. Miletić, A.; Pavlić, B.; Ristić, I.; Zeković, Z.; Pilić, B. Encapsulation of fatty oils into electrospun nanofibers for cosmetic products with antioxidant activity. *Appl. Sci.* **2019**, *9*, 2955. [CrossRef]
20. Zanin, M.H.A.; Cerize, N.N.; de Oliveira, A.M. Production of nanofibers by electrospinning technology: Overview and application in cosmetics. In *Nanocosmetics and Nanomedicines New Approaches for Skin Care*; Springer: Berlin/Heidelberg, Germany, 2011; pp. 311–332.
21. Akduman, C.; Kumbasar, E.P.A. Electrospun polyurethane nanofibers. In *Aspects of Polyurethanes*; InTech: Rijeka, Croatia, 2017; p. 17.
22. Knoff, W.F.; Kawka, D.W. Allergen Barrier Fabric. WO2008021293A1, 21 February 2008.
23. Zhang, Y.; Zhang, Y.; Zhu, Z.; Jiao, X.; Shang, Y.; Wen, Y. Encapsulation of thymol in biodegradable nanofiber via coaxial eletrospinning and applications in fruit preservation. *J. Agric. Food Chem.* **2019**, *67*, 1736–1741. [CrossRef]
24. Soukasene, S.; Toft, D.J.; Moyer, T.J.; Lu, H.; Lee, H.-K.; Standley, S.M.; Cryns, V.L.; Stupp, S.I. Antitumor activity of peptide amphiphile nanofiber-encapsulated camptothecin. *ACS Nano* **2011**, *5*, 9113–9121. [CrossRef]

25. Karagoz, S.; Kiremitler, N.B.; Sarp, G.; Pekdemir, S.; Salem, S.; Goksu, A.G.; Onses, M.S.; Sozdutmaz, I.; Sahmetlioglu, E.; Ozkara, E.S.; et al. Antibacterial, antiviral, and self-cleaning mats with sensing capabilities based on electrospun nanofibers decorated with ZnO nanorods and Ag nanoparticles for protective clothing applications. *ACS Appl. Mater. Interfaces* **2021**, *13*, 5678–5690. [CrossRef]
26. Fantini, D.; Costa, L. Dye, fluorophores and pigment coloration of nanofibers produced by electrospinning. *Polym. Adv. Technol.* **2009**, *20*, 111–121. [CrossRef]
27. Wen, X.; Xiong, J.; Lei, S.; Wang, L.; Qin, X. Diameter refinement of electrospun nanofibers: From mechanism, strategies to applications. *Adv. Fiber Mater.* **2021**, *4*, 145–161. [CrossRef]
28. Tilley, R.J. *Colour and the Optical Properties of Materials*; John Wiley & Sons: Hoboken, NJ, USA, 2020.
29. Yan, X.; You, M.-H.; Lou, T.; Yu, M.; Zhang, J.-C.; Gong, M.-G.; Lv, F.-Y.; Huang, Y.-Y. Colorful hydrophobic poly (vinyl butyral)/cationic dye fibrous membranes via a colored solution electrospinning process. *Nanoscale Res. Lett.* **2016**, *11*, 540. [CrossRef]
30. Fadil, F.; Adli, F.A.; Affandi ND, N.; Harun, A.M.; Alam, M.K. Dope-dyeing of polyvinyl alcohol (PVA) nanofibres with remazol yellow FG. *Polymers* **2020**, *12*, 3043. [CrossRef]
31. Khatri, Z.; Mayakrishnan, G.; Hirata, Y.; Wei, K.; Kim, I.S. Cationic-cellulose nanofibers: Preparation and dyeability with anionic reactive dyes for apparel application. *Carbohydr. Polym.* **2013**, *91*, 434–443. [CrossRef] [PubMed]
32. Khatri, M.; Ahmed, F.; Jatoi, A.W.; Mahar, R.B.; Khatri, Z.; Kim, I.S. Ultrasonic dyeing of cellulose nanofibers. *Ultrason. Sonochemistry* **2016**, *31*, 350–354. [CrossRef] [PubMed]
33. Khatri, M.; Ahmed, F.; Shaikh, I.; Phan, D.N.; Khan, Q.; Khatri, Z.; Lee, H.; Kim, I.S. Dyeing and characterization of regenerated cellulose nanofibers with vat dyes. *Carbohydr. Polym.* **2017**, *174*, 443–449. [CrossRef]
34. Mínguez-García, D.; Breve, N.; Capablanca, L.; Bonet-Aracil, M.; Díaz-García, P.; Gisbert-Payá, J. Liquid oil trapped inside PVA electrospun microcapsules. *Polymers* **2022**, *14*, 5242. [CrossRef]
35. Mhatre, S.; Vivacqua, V.; Ghadiri, M.; Abdullah, A.M.; Al-Marri, M.J.; Hassanpour, A.; Hewakandamby, B.; Azzopardi, B.; Kermani, B. Electrostatic phase separation: A review. *Chem. Eng. Res. Des.* **2015**, *96*, 177–195. [CrossRef]
36. Yang, D.; Wu, H.; Sun, H.; He, L.; Guo, Y. Ultra-high frequency and Self-adaptive voltage technology for water separation from oil emulsion. *Sep. Purif. Technol.* **2021**, *279*, 119732. [CrossRef]
37. Popa, R.M.; Fetea, F.; Socaciu, C. ATR-FTIR-MIR Spectrometry and Pattern Recognition of Bioactive Volatiles in Oily versus Microencapsulated Food Supplements: Authenticity, Quality, and Stability. *Molecules* **2021**, *26*, 4837. [CrossRef] [PubMed]
38. Pirzada, T.; Arvidson, S.A.; Saquing, C.D.; Shah, S.S.; Khan, S.A. Hybrid silica–PVA nanofibers via sol–gel electrospinning. *Langmuir* **2012**, *28*, 5834–5844. [CrossRef] [PubMed]
39. Bartošová, A.; Blinová, L.; Sirotiak, M.; Michalíková, A. Usage of FTIR-ATR as non-destructive analysis of selected toxic dyes. *Res. Pap. Fac. Mater. Sci. Technol. Slovak Univ. Technol.* **2017**, *25*, 103–111. [CrossRef]
40. He, Z.; Huang, C.; Wang, Q.; Jiang, Z.; Chen, J.; Song, S. Preparation of a praseodymium modified Ti/SnO2-Sb/PbO2 electrode and its application in the anodic degradation of the azo dye acid black 194. *Int. J. Electrochem. Sci.* **2011**, *6*, 4341. [CrossRef]
41. Dhanve, R.S.; Kalyani, D.C.; Phugare, S.S.; Jadhav, J.P. Coordinate action of exiguobacterial oxidoreductive enzymes in biodegradation of reactive yellow 84A dye. *Biodegradation* **2009**, *20*, 245–255. [CrossRef]
42. Mitter, E.K.; Corso, C.R. FT-IR analysis of acid black dye biodegradation using Saccharomyces cerevisiae immobilized with treated sugarcane bagasse. *Water Air Soil Pollut.* **2013**, *224*, 1607. [CrossRef]
43. Ciko, L.; Andoni, A.; Ylli, F.; Plaku, E.; Taraj, K.; Çomo, A. Extraction of essential oil from albanian salvia officinalis L. and its characterization by FTIR Spectroscopy. *Asian J. Chem.* **2016**, *28*, 1401. [CrossRef]
44. Çallıoğlu, F.C.; Güler, H.K.; Çetin, E.S. Emulsion electrospinning of bicomponent poly (vinyl pyrrolidone)/gelatin nanofibers with thyme essential oil. *Mater. Res. Express* **2019**, *6*, 125013. [CrossRef]
45. Carvalho, R.A.; de Oliveira AC, S.; Santos, T.A.; Dias, M.V.; Yoshida, M.I.; Borges, S.V. WPI and cellulose nanofibres bionanocomposites: Effect of thyme essential oil on the morphological, mechanical, barrier and optical properties. *J. Polym. Environ.* **2020**, *28*, 231–241. [CrossRef]
46. Gurav, R.; Bhatia, S.K.; Choi, T.R.; Kim, H.J.; Choi, Y.K.; Lee, H.J.; Ham, S.; Cho, J.Y.; Kim, S.H.; Lee, S.H. Adsorptive removal of synthetic plastic components bisphenol-A and solvent black-3 dye from single and binary solutions using pristine pinecone biochar. *Chemosphere* **2022**, *296*, 134034. [CrossRef] [PubMed]

MDPI AG
Grosspeteranlage 5
4052 Basel
Switzerland
Tel.: +41 61 683 77 34

Polymers Editorial Office
E-mail: polymers@mdpi.com
www.mdpi.com/journal/polymers

www.ingramcontent.com/pod-product-compliance
Lightning Source LLC
Chambersburg PA
CBHW041341210926
51850CB00013B/2950
* 9 7 8 3 7 2 5 8 4 5 0 5 7 *